Radical Chemistry: The Fundamentals

John Perkins

Professor Emeritus, University of London

Series sponsor: **ZENECA**

ZENECA is a major international company active in four main areas of business: Pharmaceuticals, Agrochemicals and Seeds, Speciality Chemicals, and Biological Products.

ZENECA's skill and innovative ideas in organic chemistry and bioscience create products and services which improve the world's health, nutrition, environment, and quality of life.

ZENECA is committed to the support of education in chemistry and chemical engineering.

OXFORD

UNIVERSITY PRESS

OXFORD

UNIVERSITY PRESS

Great Clarendon Street, Oxford OX2 6DP

Oxford University Press is a department of the University of Oxford.
It furthers the University's objective of excellence in research, scholarship,
and education by publishing worldwide in

Oxford   New York

Athens   Auckland   Bangkok   Bogotá   Buenos Aires   Calcutta
Cape Town   Chennai   Dar es Salaam   Delhi   Florence   Hong Kong   Istanbul
Karachi   Kuala Lumpur   Madrid   Melbourne   Mexico City   Mumbai
Nairobi   Paris   Saõ Paulo   Singapore   Taipei   Tokyo   Toronto   Warsaw

with associated companies in   Berlin   Ibadan

Oxford is a registered trade mark of Oxford University Press
in the UK and in certain other countries

Published in the United States
by Oxford University Press Inc., New York

A catalogue record for this book is available from the British Library

Library of Congress Cataloging in Publication Data
(Data available)
ISBN 0 19 879289 1

Typeset by the author

Printed in Great Britain
on acid-free paper by
Bath Press, Avon

Series Editor's Foreword

Radical reactions now form an essential part of the repertoire of the synthetic chemist (see OCPs numbers 8, 31, 36 and 81 for example) and industrial, synthetic, mechanistic, physical chemical and biological aspects appear throughout all undergraduate and graduate chemistry courses. It is very appropriate therefore that a primer should be devoted to the current state of knowledge of radicals and their reactions.

Oxford Chemistry Primers have been designed to provide concise introductions to all students of chemistry and contain only the essential material that would normally be included in an 8–10 lecture course. The immensely readable and pedagogical description of radicals and their reactions and applications in this Primer will be of interest to apprentice and master chemist alike.

Stephen G. Davies
Dyson Perrins Laboratory
University of Oxford

Preface

Radical reactions get scant attention in most undergraduate chemistry courses. But their importance in industrial chemistry, first properly appreciated with the drive towards synthetic substitutes for natural rubber during World War II, is enormous. If readers are deterred by polymers, they should recognise that the past twenty years have seen radical chemistry take a leading role in the theatre of the synthetic chemist. Furthermore, new examples of the importance of radical processes in biochemistry—as intermediates in essential metabolic processes, and as critical mediators of a multitude of diseases—are being established, or claimed, with increasing frequency.

Earlier volumes in this series gave a glimpse of some of the applications of radicals in synthesis. The present book sets out to paint in the essential background: the ground-rules, and the principles which underlie them, are presented here in a manner designed to give a balanced perspective on modern radical chemistry. Throughout the text, these fundamental principles are illustrated with examples from many areas of contemporary importance. It is hoped that the whole may constitute a foundation upon which the interested reader can build, enabling him or her to dip into the primary literature with assurance.

2000   M.J.P

Contents

1 Introduction

In chemistry, the first recorded use of the term 'radical' appears to have been by Lavoisier in the eighteenth century. This was to designate some poly-atomic fragment of a molecule which remains unchanged during one or more chemical transformations. In these terms, a modern example of an organic radical is ethyl, which is unaltered throughout the sequence of changes: $C_2H_5OH \rightarrow C_2H_5I \rightarrow C_2H_5CN \rightarrow C_2H_5CO_2H$.

When, in 1849, Frankland treated iodoethane with zinc and obtained zinc iodide and a gas, it was for a short time believed that the gas was the *free* ethyl radical, C_2H_5, containing trivalent carbon. However, as we now know, this chemistry gives rise to a mixture of gases, principally butane, ethane and ethene. Very soon, the universal tetravalence of carbon became an estab-lished 'fact', so that when, in 1900, Moses Gomberg declared that the yellow colour obtained by shaking a benzene solution of chlorotriphenylmethane with mercury in the absence of oxygen was due to the triphenylmethyl radical (Scheme 1.1), his conclusion was greeted with scepticism and disbelief. But Gomberg was quite correct, and a succession of triarylmethyl free radicals followed, some of which (e.g. **1.1**) can be isolated in the monomeric form.

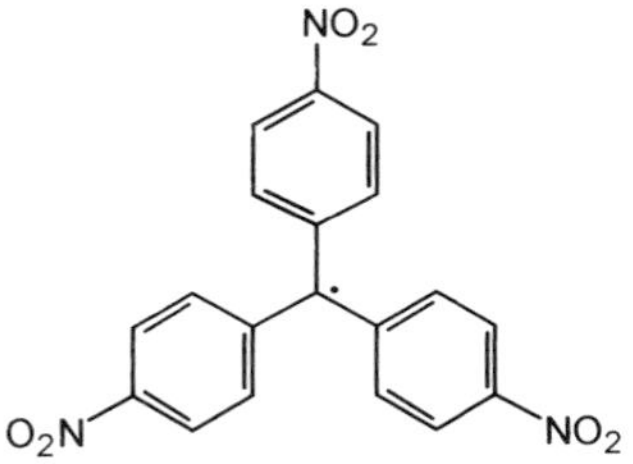

1.1 Tris-*p*-nitrophenylmethyl; a crystalline triarylmethyl

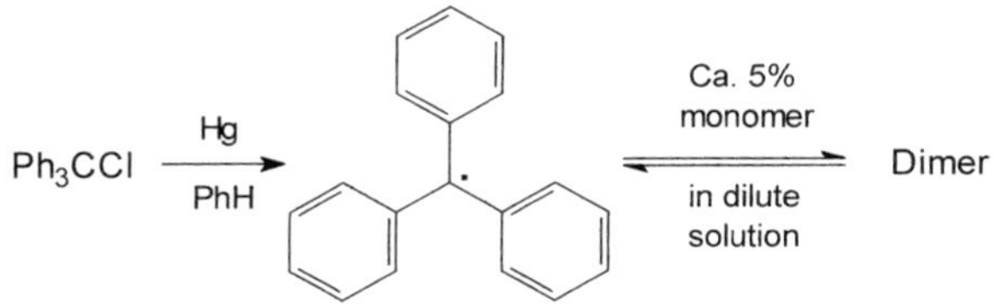

Scheme 1.1 Formation of triphenylmethyl by reduction of Ph₃CCl

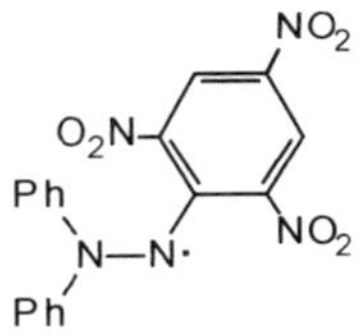

1.2 Diphenyl picryl hydrazyl (deep violet crystals)

Despite the positive identification of these carbon-centred free radicals, and of other stable species which formally contain divalent nitrogen or univalent oxygen (e.g. **1.2** - **1.4**), it was more than thirty years before there were serious suggestions that electrically neutral, short-lived species containing trivalent carbon etc. might occur as key intermediates in reactions of organic compounds in liquid solution. It is with this type of more reactive free radical that we shall principally concern ourselves in this text.

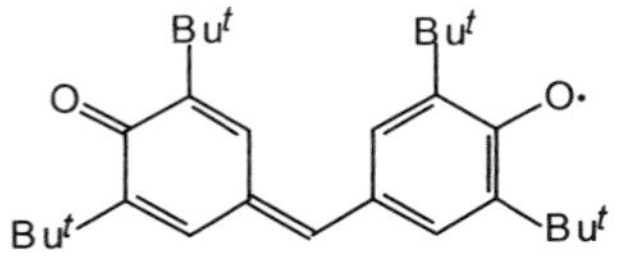

In many areas of knowledge, terminology develops, but often it does so at different rates in different sub-divisions of a subject. The modern student of chemistry is no longer expected to know the trivial (non-systematic) names of any but the simplest members of the various homologous series of aliphatic compounds. But the shift to fully systematic names is often less rapid in fields where systematisation is less important. Thus vinegar contains acetic rather than ethanoic acid, and a nurse may sniff the breath of a comatose patient in an attempt to detect not propanone but acetone for evidence of diabetes. Somewhat similar developments in terminology may

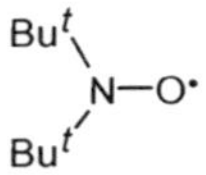

1.3 'Galvinoxyl' – a phenoxyl radical (blue crystals)

1.4 Di-*t*-butyl nitroxide (a red-brown liquid)

$PhN_2^+\ Cl^- + NaOH + C_6H_5X$

Ph—⟨⟩—X

Ph

+ ⟨⟩—X

Ph

+ ⟨⟩—X

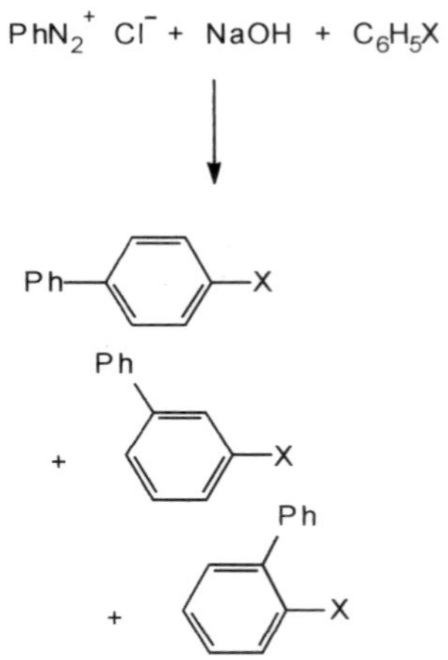

Scheme 1.2 The Gomberg phenylation reaction

Approximate isomer distribution of biphenyls from phenylation of toluene and nitrobenzene:

X	% o-	%-m-	% p-
Methyl	64	20	16
Nitro	63	10	27

The early results emphasised the formation of *ortho* and *para* products; the former, being relatively volatile, could be separated by distillation, whilst the latter could usually be separated as crystalline solids. Later experiments used gas chromatographic procedures for product analysis.

Although the chemistry described here saw the beginnings of free-radical reactions in the liquid phase, in 1924, an Austrian scientist, F. Paneth, had provided compelling evidence for the participation of methyl radicals in gas-phase chemistry. This was achieved in a series of experiments on the pyrolysis of tetramethyllead vapour.

also be found in the usage of 'radical', for, as we shall see, it is common for chemists well versed in the reactions of 'free radicals' to refer to them simply as 'radicals'. Indeed, we shall discover in Chapter 2 that the adjective 'free' may nowadays be confined to a rather specialised use by such scientists, related to some subtleties in the chemistry of these reactive species. Conversely, where the precise molecular behaviour is less relevant, such subtle distinctions have not been taken on board, and the older terminology is retained. Thus, for a cardiologist concerned about the possible benefits of antioxidants to the patient with incipient heart disease, what is important is that the antioxidant vitamins (especially C and E) may diminish 'free-radical damage' to heart tissue.

How may we define the term 'radical' (or 'free radical')? The following is a concise definition of a radical that can be found in one glossary of chemical terms: 'A molecular entity containing an unpaired electron' (the single electron is represented by the dot in displayed formulae). Some species which are embraced by this definition, but which would not normally be regarded as radicals by a practising organic chemist, include transition metal ions and alkali-metal atoms. On the other hand, halogen atoms *are* commonly accepted as coming within the definition. Perhaps this is because many halogenation reactions are amongst the most familiar of simple radical-mediated processes.

Some of the earliest evidence that electrically neutral carbon-centred radicals may be important intermediates in organic chemistry came from observations on a class of aromatic substitution reactions in which hydrogen is replaced by a phenyl group (Scheme 1.2). In reactions of monosubstituted benzenes, it became clear that there was no strong directing effect of the benzene substituent, irrespective of the electronic nature of that substituent. All three isomeric products are invariably formed in significant amounts, and the relative reactivities of benzenes with electron-donating or -withdrawing substituents (e.g. toluene and nitrobenzene) are quite similar. The conclusion that it was an *electrically neutral* phenyl radical (**1.5**) which was responsible for effecting these substitution reactions was, at the time, highly controversial. The initial report, published in 1934 by the British chemist D.H. Hey, therefore appeared under the less contentious title 'Amphoteric Aromatic Substitution', but the free-radical interpretation was quite explicit in the discussion of results.

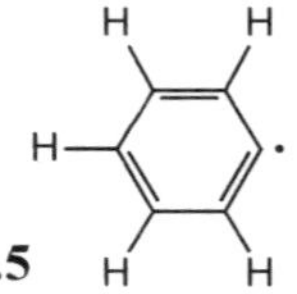

1.5

At almost the same time, M. Kharasch in Chicago was investigating the addition of HBr to alkenes. The expected Markovnikov addition product was obtained only when the reaction was observed to proceed rather slowly. On occasion, however, reaction was much faster. When this happened, the anti-Markovnikov product was obtained. The conclusion was that the slow reaction followed the ionic pathway, established for many other addition reactions; the more rapid process was traced to the presence of peroxidic impurities. This 'peroxide effect' was eventually attributed to the intervention of neutral free-radical intermediates which caused bromine atoms to be formed from the hydrogen bromide, and resulted in the sequence of reactions shown in the lower section of Scheme 1.3.

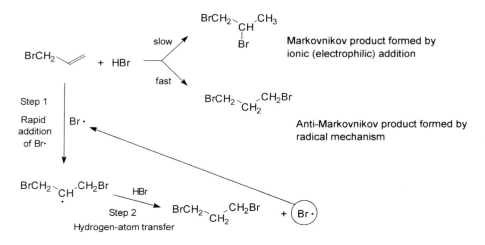

Scheme 1.3

Also in the early 1930s, the gas-phase chlorination of methane was interpreted in terms of a radical mechanism. Today, this is amongst the very first organic transformations which the student of chemistry is likely to encounter (Scheme 1.4). The liquid-phase halogenation of less volatile alkanes and cycloalkanes follows a similar mechanism.

Step 1   Cl· + CH₄ ⟶ HCl + CH₃·

Step 2   CH₃· + Cl₂ ⟶ CH₃Cl + Cl·

Scheme 1.4

In both of the above reaction schemes, the first step begins with a halogen atom and the second step produces one – which can, of course, in each case, re-enter Step 1. The repetitious character of these processes gives rise to the idea of a 'chain reaction'. And chain reactions comprise the majority of radical processes that will be examined in the following pages. In both steps of both of the above reaction schemes, one radical is replaced by another. But the reader will at once have noticed that the halogen atom must first have arisen from some other process. Furthermore, unless the chain is to continue indefinitely, there must be incursion of a further type of reaction in which radicals are destroyed. The steps in Schemes 1.3 and 1.4 are known as *chain-propagating* reactions. Processes whereby radicals are formed from non-radicals are *chain-initiating* reactions, and those in which radicals are destroyed are *chain-terminating* reactions. The several types of reaction steps involving radicals will be discussed and exemplified in Chapter 2. Only a tiny amount of initiation may be sufficient to account for bulk transformation of reactant into product. This will be determined by the relative efficiencies of propagation and termination. In some instances, the propagating steps will repeat hundreds or even thousands of times before termination intervenes. The average number of repeats is the 'chain length'.

In recent years, there have been important developments in which radical chemistry has been applied to synthesis.   For example, the complex transformation shown in Eqn 1.1 proceeds by a radical-chain mechanism with a yield exceeding 70%. Yet until as little as twenty-five years ago, the

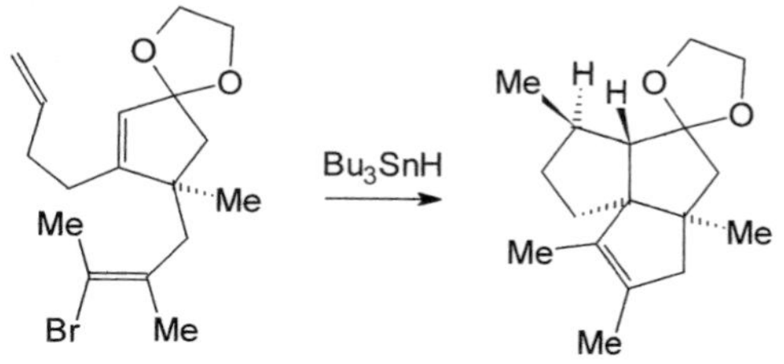

$$(1.1)$$

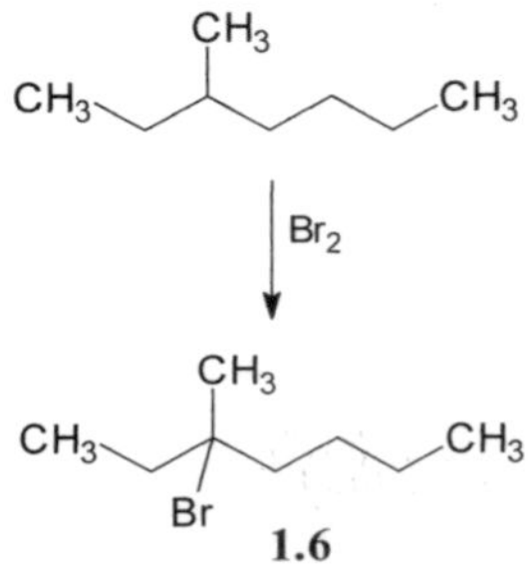

prospects for synthesis with radicals seemed, with a few notable exceptions, to be very poor. Partly this reflected a lack of quantitative data on reaction rates, but also it stemmed from a general sentiment that reactive radicals were very unselective in their behaviour. An example of this has already been noted in the case of aromatic phenylation in which *ortho, meta,* and *para* isomers are formed in roughly comparable amounts. And in chlorination of alkanes more complex than ethane, e.g. 3-methylheptane, all possible mono-chloro derivatives will be formed in appreciable quantities. In contrast, however, from the analogous bromination reaction as much as 70% of the mono-brominated product will be the single isomer, **1.6**.

There was one aspect of the chemistry of reactive radicals that was already very extensively investigated within little more than a decade of the early observations on phenylation.   This concerned polymerisation of vinyl monomers, typified by the formation of polystyrene depicted in Scheme 1.5.

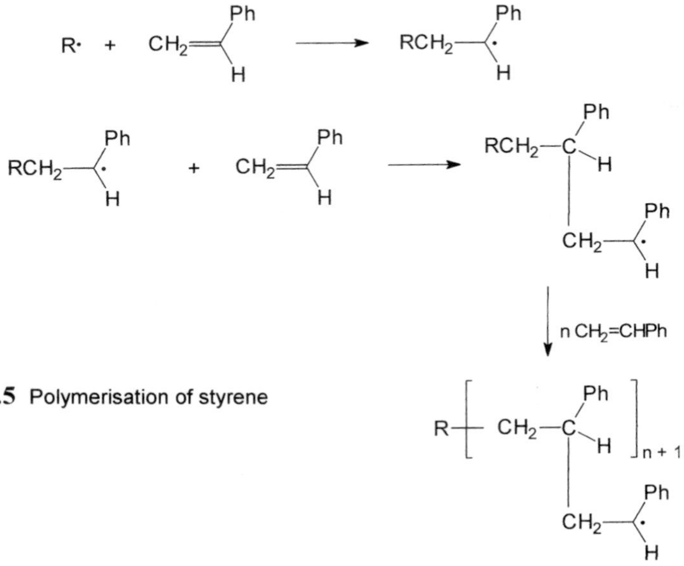

Scheme 1.5 Polymerisation of styrene

These efforts were prompted by the search, during the second world war, for a synthetic alternative to natural rubber, when supplies to the Western Allies from what is now Malaysia were no longer accessible. Although these investigations afforded a wealth of quantitative information, the formation of rubber-like or glassy polymers had little impact on the development of this type of chemistry for the efficient synthesis of small complex molecules. But the reader might reflect for a moment on the fact that conversion of monomer into polymer frequently approaches 100%. Like the example of bromination given above, this discloses the potential for high yields, at least in some radical reactions.

It may not be immediately obvious why the bromination reaction should be so much more selective than chlorination. Even when it is appreciated that the chlorine atom is generally much more reactive than is bromine, the question arises: does it necessarily follow that greater reactivity is paralleled by diminished selectivity? This inverse correlation will be examined in some detail in Chapter 3. For the present, it is sufficient to observe that the greater reactivity of the chlorine atom is a reflection of the greater strength of the bond it forms with hydrogen (i.e. the bond dissociation energy of HCl) (431 kJ mol^{-1}) compared to that formed by a bromine atom (BDE of HBr = 366 kJ mol^{-1}).

The fission of HBr or HCl into atoms is referred to as 'homolysis'. This indicates that there is an even split of the molecule in the sense that each atom acquires one of the two bonding electrons. It often comes as a surprise that this is a much easier process than is the far more familiar 'heterolysis' to proton and halide ion, in which both electrons are retained by the halogen atom. For example, the ionisation of HCl into H$^+$ and Cl$^-$ requires 1393 kJ mol^{-1}. But this, of course, refers to the gas phase. The energies of solvation of the ions is so great in water that ionisation is effectively complete. One of the simplifying features of radical chemistry is that we are usually dealing with electrically neutral species. As a result of this, solvent effects on reaction rate are often negligibly small, even in the extreme circumstance of transfer from hydrocarbon solvent to water, although radical attack on some solvents may be a complicating factor. In this respect, it is worth mentioning that water is essentially inert to radical attack. This reflects the great strength of the O–H bonds in water (BDE of HO–H = 498 kJ mol^{-1}), and their consequent lack of homolytic reactivity. It stands in marked contrast to the complications which may arise when water or other hydroxylic species are present during ionic reactions. Instead, in radical chemistry, it is molecular oxygen which must usually be excluded. Molecular oxygen is itself a radical, or, more correctly, a 'biradical', sometimes represented as ·O–O·. Not surprisingly, therefore, it, too, rapidly intercepts organic radicals of many kinds, ultimately leading to products of 'autoxidation'. Typically, in the work mentioned above, in which Gomberg generated triphenylmethyl, air was not excluded from the first experiments, and a colourless crystalline solid $C_{38}H_{30}O_2$ was isolated. This proved to be the organic peroxide, **1.7**.

As with molecular oxygen, the presence of unpaired electrons renders organic free radicals paramagnetic. In principle, this property is detectable by conventional magnetic techniques, but in practice this is possible only with high concentrations of relatively stable radicals. However, electron spin resonance spectroscopy (e.s.r.), which will be outlined in Chapter 4, is a

H—Cl

heterolysis homolysis

H$^+$ + Cl$^-$ H· + Cl·

1.7

Representations of the radical cation and radical anion of anthracene, formed from the hydrocarbon by one-electron oxidation and one-electron reduction respectively.

technique which is very much more sensitive to the paramagnetism of molecules with unpaired electrons.

In the electronic ground state of a radical, the unpaired electron will normally be found in the highest energy occupied molecular orbital. Because there is but one electron in this orbital, it is commonly referred to as the Singly Occupied Molecular Orbital (SOMO). On closer examination it will usually be found that this is a non-bonding orbital. In such circumstances, it is perhaps not surprising that radicals may be susceptible to one-electron oxidation or reduction in which the unpaired electron is removed to form a cation, or is joined by a partner to produce an anion. When molecular oxygen undergoes one-electron reduction, the superoxide 'radical anion', O_2^-, is formed. This species is known in the inorganic salt, KO_2, but it is also a very important participant in key biochemical processes. One-electron oxidations and reductions are also possible with ordinary diamagnetic organic molecules, i.e. those in which all electrons are paired. Some of the most familiar examples are found with molecules having extended π-electron conjugation, for which an electron is rather easily removed from the (π) HOMO, or inserted into the (π^*) LUMO. Typically, anthracene in dry, oxygen-free tetrahydrofuran can be reduced by sodium metal to give the corresponding radical anion, whilst electrochemical oxidation gives the radical cation. We shall examine some aspects of radical ion chemistry in Chapter 6.

2 The elementary reaction steps

2.1 Introduction

The organic chemist is interested in reactions of compounds which yield products that may be isolated and purified. Where such transformations involve radicals, it is convenient to learn about them by first acquiring some familiarity with the individual reaction steps in which the radicals may participate. Thus, individual steps in the chain reaction represented in Scheme 1.4 involve atom-transfer, whilst steps in Schemes 1.3 and 1.5 demonstrate radical addition to double bonds. In this chapter, we shall concentrate particularly on the variety of these elementary 'unit steps', such as radical addition and atom-transfer, in which radicals may participate. In doing so, it is important that we try not to lose sight of the bulk transformations in which such steps may play a part.

We shall begin our discussion of these essential unit steps with interactions between two radicals. Crucially, these halt chain reactions by giving non-radical products.

2.2 Radical-radical reactions

When two radicals interact to form non-radical products, this is chain termination. Two processes are possible: radical coupling, and disproportionation. These are illustrated here by the interaction between ethyl and methyl. Note that the term 'dimerisation' should strictly be restricted to coupling of two identical radicals.

$$CH_3\cdot + C_2H_5\cdot$$

$CH_3CH_2CH_3$ — radical coupling

$CH_4 + CH_2=CH_2$ — disproportionation

$$CH_3\cdot + CH_3\cdot$$

$$C_2H_6$$

An example of 'dimerisation'

For the vast majority of reactive or even moderately reactive radicals, there is at most a tiny activation barrier to these processes. The result is that their rates are limited only by diffusion through the liquid medium. In a mobile solvent, such as cyclohexane or water, this implies a bimolecular rate constant of between 10^9 and 10^{10} M^{-1} sec^{-1}. An interesting and very important consequence is that dimerisation of, for example, resonance-stabilised benzyl radicals (**2.1**) (overleaf) occurs at a very similar rate to that of methyl radicals. Even cross-coupling of e.g. benzyl with a stable radical such as di-t-butyl nitroxide (**1.4**) will occur almost as rapidly (ca 5×10^8 M^{-1} sec^{-1}).

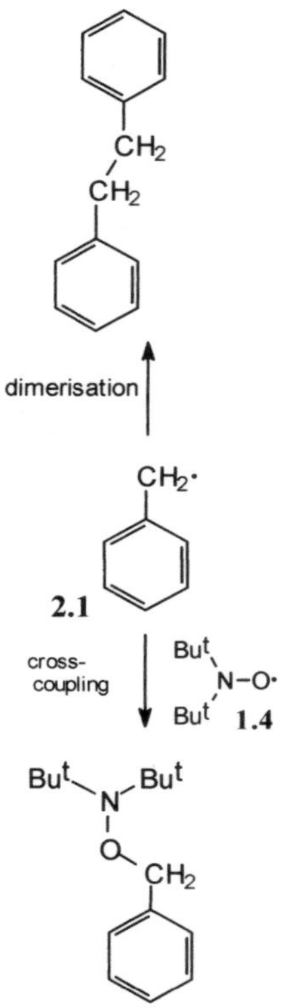

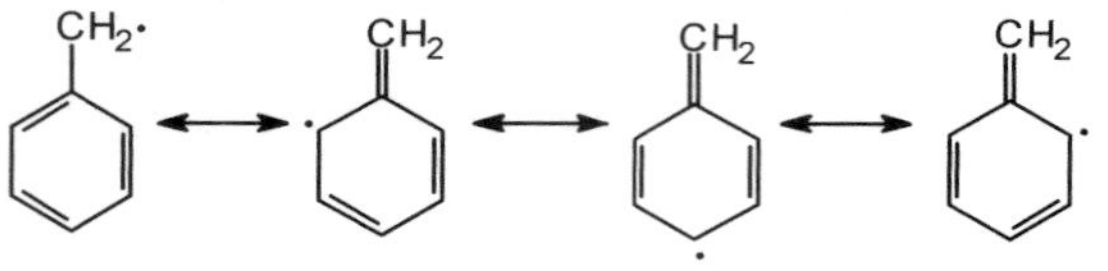

2.1 The resonance stabilised benzyl radical

Although the emphasis here is on chain termination, when a radical is generated in a solvent towards which it is unreactive, then dimerisation can become a major product-forming process. A classical example is the Kolbe anodic oxidation of salts of carboxylic acids (Scheme 2.1). Removal of an electron from carboxylate gives the carboxyl radical, but this is unstable and loses carbon dioxide. The resulting alkyl radical then gives modest yields

$$RCO_2^- \xrightarrow{-e} RCO_2\cdot$$

(at anode in electrolysis apparatus)

$$RCO_2\cdot \longrightarrow R\cdot + CO_2$$

Scheme 2.1

$$2R\cdot \longrightarrow R{-}R$$

of dimer. Another long-established example is the interesting oxidation of monohydric phenols to give quinones, using the inorganic nitroxide radical (**2.2**), known as Fremy's radical (or Fremy's salt). In this process, the Teuber Reaction, the phenolic oxygen is removed by one molecule of Fremy's radical, whereupon cross-coupling occurs between the derived phenoxyl radical and a second molecule of Fremy's radical. The sequence is completed by decomposition of the product to form a quinone (Scheme 2.2).

Note that the oxygen-centred radical formed by oxidation of a phenol is referred to as a phenoxyl. Some authors use 'phenoxy'. The '-yl' ending is generally preferred here, *cf.* alkoxy, acyloxy, etc. An exception is found with nitroxides, although for these, 'aminyloxy' or 'aminoxyl' are sometimes used.

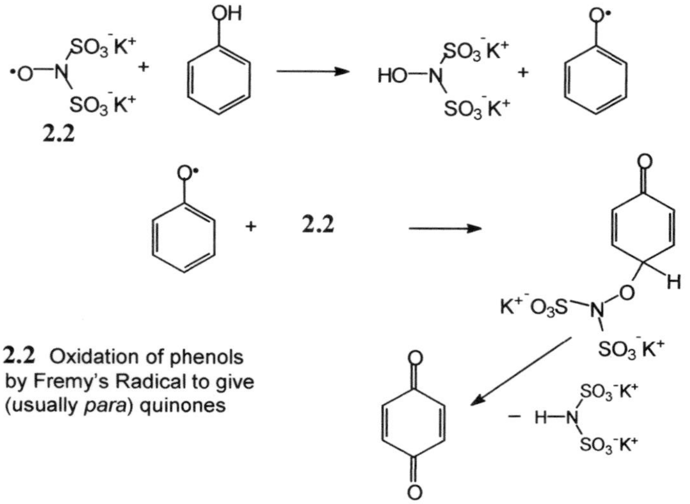

Scheme 2.2 Oxidation of phenols by Fremy's Radical to give (usually *para*) quinones

One special case of radical coupling, which also occurs with a diffusion limited rate, is that of a carbon-centred radical with molecular oxygen, already referred to in Chapter 1. Since oxygen is a biradical, the initial product is a peroxyl radical, and this *is* capable of propagating a radical chain. Radical-chain oxidation of an organic species, R–H, by molecular oxygen is represented in Scheme 2.3. This type of oxidation sequence is very important, and we shall return to it again. It contributes to many oxidative processes, including the 'drying' of oil-based paints, the oxidative degradation of lubricating oils, and the modification of certain natural lipids in living organisms. (See also Chapter 5).

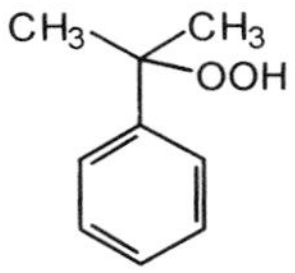

Cumene hydroperoxide, obtained by autoxidation of the hydrocarbon, cumene (isopropylbenzene), was once an important commercial precursor to phenol, from which it is formed, together with acetone, by an acid-catalysed rearrangement. Can you write a mechanism for this rearrangement?

Scheme 2.3 Chain-propagating steps for an autoxidation reaction. The initial product is a hydroperoxide, ROOH. This process is especially important when the R–H bond is weak.

2.3   Radical-molecule reactions

A second class of bimolecular process embraces the vast majority of chain-propagating events in which a radical interacts with a non-radical. These are typified by atom-transfer – especially hydrogen atom-transfer – reactions, and by addition reactions, both of which we have already encountered.

One organic functional-group interconversion in which atom transfer is involved, and which has assumed considerable synthetic importance in recent years, is the reduction of halogen compounds, R–X (especially aliphatic and alicyclic bromides and iodides), by organotin hydrides. The overall reaction and chain-propagating steps are set out in Scheme 2.4. Qualitatively, one may say that the success of this reaction depends on the relative weakness of the Sn–H bond, and the high affinity of tin for halogen.

Overall reaction:

$$Bu_3SnH \ + \ RX \ \longrightarrow \ RH \ + \ Bu_3SnX$$

Chain-propagating steps:

$$Bu_3SnH \ + \ R\cdot \ \longrightarrow \ RH \ + \ Bu_3Sn\cdot$$

$$Bu_3Sn\cdot \ + \ RX \ \longrightarrow \ Bu_3SnX \ + \ R\cdot$$

Scheme 2.4

An important variant of the reduction of organic halogen compounds by tin hydrides is found in the reduction of organochalcogen componds. This has been especially developed in the case of selenophenyl derivatives, RSePh

(Eqn 2.1), but succeeds with sulphur and tellurium analogues. The reaction

$$Bu_3SnH + RSePh \longrightarrow RH + Bu_3SnSePh \qquad (2.1)$$

depends on homolytic displacement of R· from selenium (the alkyl–Se bond is weaker than aryl–Se) in what has been referred to as an S_H2 (substitution homolytic bimolecular) reaction.

A logical development of the mechanistic chemist's 'curly-arrow' notation has been adopted by radical chemists in which 'fish-hooks' are used to represent single-electron movement. Eqn 2.2 uses these to show electron reorganisation in the S_H2 attack on selenium. Whilst this provides a full picture of the rebonding which is occurring, it is commonly abbreviated as in Eqn 2.3. Here, single-electron flow is shown in only one direction, the accompanying reverse flow being understood. It also reveals the obvious mechanistic similarity to an S_N2 process. Of course, atom-transfer processes also fall into the S_H2 category (Eqn 2.4), although the designation 'atom transfer' is usually preferred.

The occurrence of S_H2 reactions at carbon is rare, and should be offered as a mechanistic rationalisation only in exceptional circumstances. The best known examples are found in reactions of highly strained species, e.g. :

$$Bu_3Sn\cdot \quad \overset{}{\underset{Ph}{Se}}\!\!-\!R \longrightarrow Bu_3SnSePh + R\cdot \qquad (2.2)$$

$$Bu_3Sn\cdot \quad \overset{}{\underset{Ph}{Se}}\!\!-\!R \longrightarrow Bu_3SnSePh + R\cdot \qquad (2.3)$$

$$X\cdot \quad H\!-\!R \longrightarrow X\!-\!H + R\cdot \qquad (2.4)$$

Addition reactions, whilst extensively documented in the context of polymer chemistry, need not lead to polymeric products. The example of HBr addition has been mentioned (p. 3). Even readily polymerised substrates such as styrene may, in some circumstances, give good yields of 1:1 addition products. One example is encountered in the reaction between styrene and carbon tetrabromide, where bromine-atom transfer from carbon tetrabromide to the benzylic radical (**2.3**) is sufficiently rapid that the addition to a further molecule of styrene fails to compete effectively except when the concen-

$$\cdot CBr_3 + PhCH\!=\!CH_2 \longrightarrow \underset{\textbf{2.3}}{Ph\dot{C}H\text{-}CH_2CBr_3}$$

$$\textbf{2.3} + CBr_4 \xrightarrow{k_{CBr_4}} PhCHBrCH_2CBr_3 + \cdot CBr_3$$

$$Ph\dot{C}H\text{-}CH_2CBr_3 + PhCH\!=\!CH_2 \xrightarrow{k_{styrene}} \underset{\underset{CH_2\dot{C}HPh}{|}}{PhCHCH_2CBr_3}$$

$$k_{CBr_4} \gg k_{styrene}$$

tration of CBr$_4$ is very low. Under these latter circumstances, short chain-length polymers may form before the growing polymer is intercepted by the CBr$_4$. This interception of the growing polymer is an example of 'chain transfer'. The resulting radical, in this case ·CBr$_3$, may then initiate the growth of a further chain. Short chain-length polymers, incorporating only a few monomer units, are referred to as 'telomers' or 'oligomers'.

If the growing polymer is intercepted by a molecule of a second vinyl monomer, the resulting radical may then react with further monomer units. In this case, we have an example of radical-chain 'copolymerisation'. But this phenomenon is far from straightforward. For example, initiating polymerisation of an equimolar mixture of two vinyl monomers – let us designate them MA and MB – may have a variety of consequences ranging from a mixture of poly(MA) with poly(MB), to a single copolymer with random sequences of MA and MB, or, as we shall see in Chapter 3, a copolymer with almost perfect alternation of MA and MB along the polymer chain. This complexity results from the wide range of rate constants exhibited by any growing polymer radical towards different monomers; factors come into play which dictate that the patterns of reactivity of different growing chain radicals, e.g. poly(MA)· and poly(MB)· may be surprisingly different.

There is now a wealth of documentation on radical additions to most of the different types of multiple bonds commonly found in organic molecules, and many of these have been elegantly incorporated into modern synthetic methodology, although a consequence of the great strength of the carbonyl double bond is that the reverse process (see below) is far more common. Furthermore, only the most reactive radicals will disrupt the π-system of a simple benzene ring. One that will is the phenyl radical, already encountered in the brief historical introduction in Chapter 1. The unpaired electron in phenyl is localised in an sp^2 orbital on carbon. When this radical adds to benzene, it forms a resonance stabilised phenylcyclohexadienyl radical (**2.4**), in which the unpaired electron is in a π orbital and is delocalised over five carbon atoms. Nevertheless, the resonance stabilisation of this adduct is appreciably less than that of benzene itself. Whilst there is an obvious similarity between **2.4** and the cationic 'Wheland Intermediate' of e.g. electrophilic nitration (**2.5**), onward conversion of **2.4** to biphenyl is far less straightforward than is that of the Wheland intermediate into nitrobenzene – a point to which we shall return later (p. 27).

We have noted that the majority of radical-radical reactions occur at rates determined only by diffusion through the solvent. For radical-molecule reactions there is no simple generalisation of this kind. The fastest, involving very reactive radicals such as hydroxyl (HO·), may also occur at, or close to, the encounter- (diffusion-)controlled limit, whereas others will be undetectably slow. In between, commonly with rate constants varying from *c.* $10 - 10^6$ M^{-1} sec^{-1}, are the bimolecular propagating steps of various chain reactions. Representative examples of rate constants for radical-molecule reactions are given in the Appendix, and an outline of some of the techniques whereby these numbers have been determined can be found in Chapter 4.

Copolymerisation of styrene and acrylonitrile

Double bond type	Approx. b.d.e. (kJ M^{-1})
C=C (alkene)	600
C=O (ketone)	750

2.4

2.5

2.4   Unimolecular reactions

Whilst a radical-molecule reaction will normally lead to a new radical capable of carrying on the reaction chain, there are unimolecular processes which may also be described as chain-propagating.   These include rearrangements and fragmentation reactions, exemplified in Eqns 2.5 and 2.6 respectively.

$$\text{Me}-\overset{\overset{\displaystyle Ph}{|}}{\underset{\underset{\displaystyle Me}{|}}{C}}-CH_2\cdot \quad\longrightarrow\quad \text{Me}-\overset{\overset{\displaystyle \cdot}{|}}{\underset{\underset{\displaystyle Me}{|}}{C}}-\overset{\displaystyle Ph}{CH_2} \tag{2.5}$$

$$\underset{\underset{\displaystyle Me}{|}}{\overset{\overset{\displaystyle PhCH_2}{\diagdown}}{\text{MeO}-C}}-O\cdot \quad\longrightarrow\quad PhCH_2\cdot \;+\; \overset{\displaystyle MeO}{\underset{\displaystyle Me}{\diagup}}C=O \tag{2.6}$$

2.6

Radical Rearrangements. Any reader with even a passing acquaintance with the chemistry of carbocations will be familiar with the tendency for these species to reorganise themselves into more stable structures.   A now classic example is the formation of $Me_3C^+\,SbF_6^-$ from *any* isomer of C_4H_9F when the latter is dissolved in liquid sulphur dioxide containing SbF_5.   Although, like tertiary cations, tertiary radicals are more stable than are their primary or secondary counterparts (see Chapter 3), simple aliphatic radicals show little tendency to undergo rearrangement.   Qualitatively, it can be considered that the extra electron 'gets in the way', so that a rearrangement transition state such as **2.6** is energetically inaccessible, at least during the fleeting lifetime of a reactive radical.   Where, however, there is a low-lying orbital capable of accommodating the unpaired electron, then 1,2-shifts can be observed.   This occurs when the migrating group has accessible *d*-orbitals, e.g. Eqn 2.7, or a conveniently located π-system, as depicted in Eqn 2.8.

$$\tag{2.7}$$

$$\tag{2.8}$$

Possibly because 1,2-shifts are relatively rare in radical chemistry, intramolecular analogues of atom transfer and of radical addition are commonly classified as rearrangements. The most frequently encountered examples of intramolecular hydrogen-atom transfer involve 1,5-shifts, i.e. six-membered-ring transition states, exemplified in Eqn 2.9.

$$(2.9)$$

One application of this, which, when it was first reported some thirty years ago marked a giant stride in the application of radical chemistry in organic synthesis, was D.H.R. Barton's functionalisation of the C(18) methyl group in the steroid skeleton. This permitted the first synthesis of the hormone aldosterone (**2.7**).

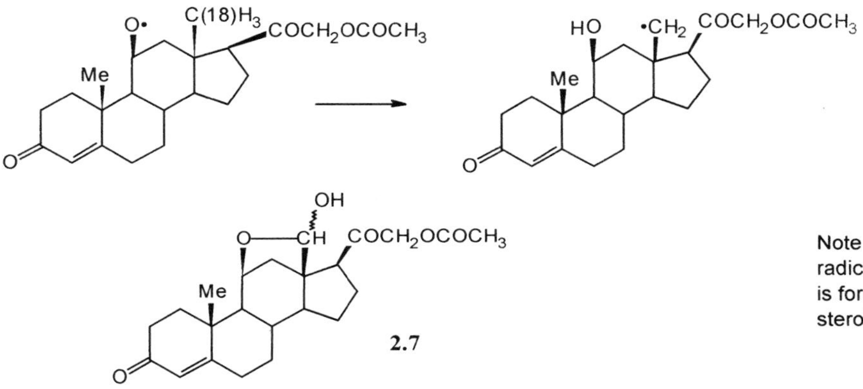

2.7

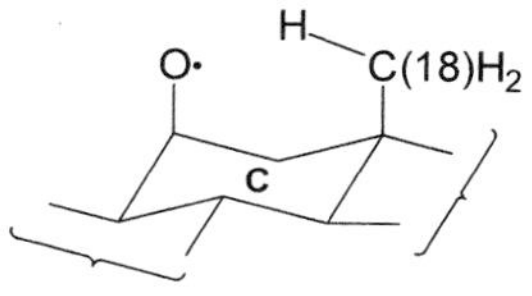

Note the proximity of the oxygen radical and the C(18) methyl, which is forced by the geometry of the steroid C-ring.

By far the best documented examples of intramolecular *addition* are those of 5-hexenyl radicals (**2.8**) and their analogues. These generally cyclise to give five-membered rings, a fact which has been extensively deployed in synthetic chemistry as a means of generating cyclopentanes, sometimes in so-called 'tandem' ring-closures where more than one cyclisation occurs. An elegant example is shown in Scheme 2.5, in which all of the propagating steps responsible for the transformation displayed in Eqn 1.1 (p. 4) have been

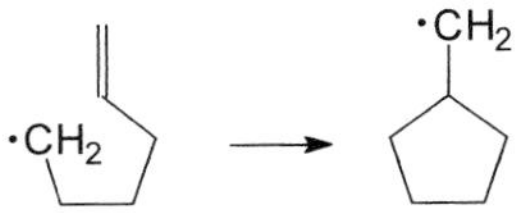

2.8

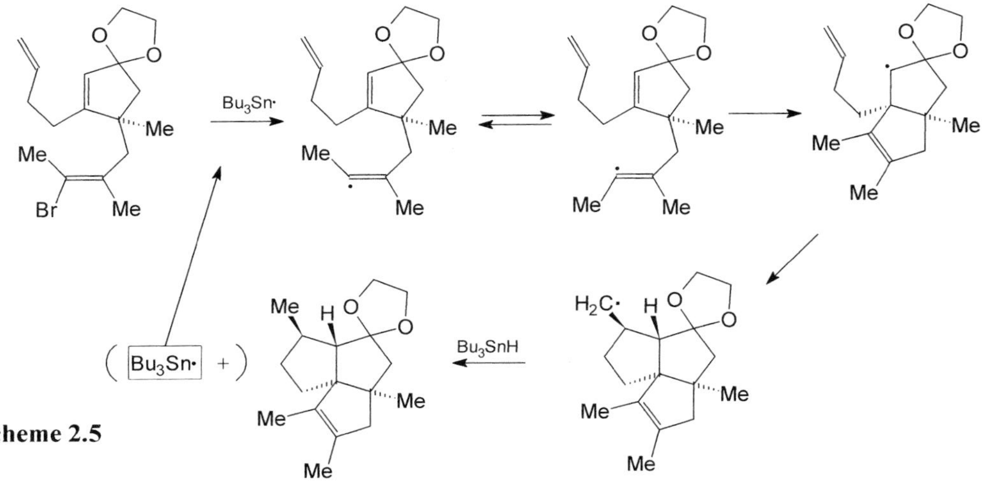

Scheme 2.5

For the thoughtful reader, the hexenyl radical cyclisation, illustrated in Scheme 2.5, may raise some important questions. Why, for example, should the hexenyl radicals cyclise as indicated, rather than to give (more stable) cyclohexyl radicals? And why should the cyclisation reactions be so much more efficient than interception of uncyclised intermediates by tin hydride? (The bimolecular addition of a primary alkyl radical to propene is about 100 times slower than its reaction with tributyltin hydride.) These subtleties comprise important topics to be addressed in Chapter 3.

Note also that the initial vinyl radical is configurationally unstable: the *cis-trans* equilibration of vinyl radicals is extremely rapid.

set out in order to indicate how tin hydride chemistry is much more than just a method of reducing organic halogen compounds.

There is one type of radical rearrangement which merits special attention, both because of the mechanistic challenge which it has posed to numerous investigators since its initial discovery some 30 years ago, and also for its occurrence during radical reactions of carbohydrate derivatives, some of which have found synthetic importance. This is the 1,2-shift of an acyloxyl group. The original example is shown in Eqn 2.10. In this case, ^{18}O-labelling has revealed that there is essentially clean conversion of carbonyl oxygen in the unrearranged radical (**2.9**) into ether oxygen in the product radical (**2.11**), consistent with the intermediacy of the cyclic radical (**2.10**). However, although **2.10** does ring open to **2.11**, it has been shown that *it does so more slowly than* **2.9** *undergoes rearrangement.* Therefore **2.10** cannot be on the pathway from **2.9**. More recent investigations have also found examples in which the acyloxyl group migrates *without* interconversion of ether and carbonyl oxygens. A full discussion of this intriguing system is beyond the scope of an introductory text. The interested reader is referred to the reading list in the Appendix.

$$(2.10)$$

Fragmentation reactions. In the Section on radical-molecule reactions it was indicated that additions to most types of unsaturated structure have been documented. Where, however, the π-bonding is exceptionally strong, it is not unusual to observe the reverse process, usually referred to as fragmentation. The decarboxylation of acyloxyl radicals (RCO$_2$·) in the Kolbe reaction (p. 8) is an example of this. Other representative examples are shown in Eqns 2.11–2.13.

$$(2.11)$$

$$(2.12)$$

$$(2.13)$$

Intramolecular fragmentations, of which that of **2.12** is illustrative, are generally classified with rearrangements. A very important case, to which we shall return in Chapter 4, is the (reversible) ring-opening of the cyclopropylcarbinyl radical (Eqn 2.14).

$$ \text{(2.14)} $$

2.5 Radical-forming processes

Radicals which it is possible to isolate lend themselves to formation by conventional chemical methods, usually of oxidation or reduction. Thus the examples of carbon-, nitrogen-, and oxygen-centred radicals mentioned in Chapter 1 may be prepared by such methods. Of these radicals, only the carbon-centred triarylmethyl (**1.1**) will react rapidly with molecular oxygen, although the colours of solutions of DPPH or galvinoxyl will very slowly fade in the presence of air.

Of greater concern to us must be the formation of short-lived, reactive radicals capable of initiating chain reactions. This may come about in any one of several ways.

Thermolysis. Thermolysis of weak covalent bonds presents no particular problem. For example, when di-*t*-butyl peroxide is heated above *c.* 130°, the molecule decomposes in a cleanly first-order process the activation energy of which is equated with the dissociation energy of the O–O bond. Like all peroxide bonds, this is relatively weak, a fact which may be attributed to the electrostatic repulsion between the lone-pair electrons on the two oxygen atoms.

Not all peroxides are as kinetically well behaved as is di-*t*-butyl peroxide. This is because, in addition to the spontaneous homolysis of the peroxide bond, there is sometimes an 'induced' component to the decomposition, in which radicals generated following the initial unimolecular homolysis may subsequently interact with a second molecule of peroxide. Indeed, there is a possible explosion hazard when certain peroxides are mixed with particularly sensitive substrates. Nevertheless, careful kinetic study of closely related peroxides, in solvents where this type of induced decomposition can be demonstrated to be unimportant, reveals some interesting features. For example, the peroxyester **2.14** decomposes with a half-life of about 15 minutes at 90°C, which is some 1200 times faster than decomposition of the parent peroxyacetate (**2.13**) under the same conditions. This has been interpreted in terms of a concerted two-bond fission in the case of **2.14**,

Half-life at 130° circa. 100 h.

Thermal homolysis of di-*t*-butyl peroxide to give two *t*-butoxyl radicals.

facilitated by the formation of the resonance-stabilised benzyl radical. Supporting evidence for this comes from the experimentally determined entropies of activation for decomposition of the two peroxides, the lower value (8 J deg^{-1} mol^{-1}) for **2.14** suggesting significantly greater order in the transition state than is the case for the parent peroxyacetate ($\Delta S^{\ddagger} = 85$ J deg^{-1} mol^{-1}).

Di-*t*-butyl peroxyoxalate (**2.15**) is another peroxide which is believed to decompose by simultaneous fission, in this case of three bonds. This compound is a useful, low-temperature source of *t*-butoxyl radicals (half-life of a few minutes at 40°), but the colourless crystals are extremely shock sensitive, and the compound must be handled with great care.

In practice, the explosion hazard with many common organic peroxides is not very great. This is particularly true when the peroxide unit comprises only a relatively small fraction of the molecule. Thus dibenzoyl peroxide ('benzoyl peroxide') is a less hazardous substance than is diacetyl peroxide, although in dilute solution these compounds decompose at roughly comparable rates. Relative rates of decomposition of the peroxides discussed in this Section are listed in Table 2.1.

Members of one other class of compound are also valuable radical initiators. These are the azo-compounds, probably best known to the reader in the form of azobenzene (Ph–N=N–Ph) and the related azo dye-stuffs. Many of these aromatic azo-compounds are thermally stable below about 300°C, but certain aliphatic analogues decompose quite rapidly at temperatures below 100°C. One of the best known of these is 'AIBN', azoisobutyronitrile (**2.16**). This has a half-life in solution of about 1½ hours at 80°C, decomposing into nitrogen and two resonance-stabilised cyano-propyl radicals. Although the cyanopropyl radicals are most usually represented as being 'carbon-centred' (**2.17a**), i.e. having the unpaired electron on carbon, there is significant resonance stabilisation by contribution from the ketimine structure (**2.17b**). Evidence for this comes, *inter alia*, from the very much slower (by a factor of almost 10^6 at 80°C) decomposition of azoisobutane (Me_3C–N=N–CMe_3).

2.15

Table 2.1 Approximate relative rates of unimolecular thermolysis of selected peroxides at 80°C in a non-polar solvent.

t-BuO–OBu-*t*	0.2
MeCO₂–OBu-*t* (**2.13**)	1.0*
PhCH₂CO₂–OBu-*t* (**2.14**)	1200
(MeCO₂)₂	500
(PhCO₂)₂	200
Peroxyoxalate **2.15**	10^6

*Half-life at 80°C *c.* 600 h

Photolysis. An alternative to heat as a means of providing energy with which to cleave covalent bonds is irradiation with visible or ultraviolet light. Here, sufficient energy may be available to break rather stronger bonds than those between two oxygen atoms, though an important word of caution is in order. Absorption by any molecule of a quantum of red light of wavelength 500 nm is equivalent to imparting some 240 kJ mol^{-1} to the bulk compound. One might expect, therefore, that di-*t*-butyl peroxide, (dissociation energy 160 kJ mol^{-1}) would be decomposed by red light. It is not. The naive expectation is based on the faulty assumption that red light will be absorbed by the peroxide. But di-*t*-butyl peroxide is almost colourless.* It is, however,

---

* In large amounts a slight yellow colour is evident

cleanly broken into two butoxyl radicals by near ultraviolet light, which *is* absorbed.

All peroxides, including alkyl hydroperoxides, ROOH, are easily photolysed by ultraviolet light. In the case of hydrogen peroxide, this affords a means of generating hydroxyl radicals, HO·.

Radicals are also produced by photolysis of some azo-compounds, including AIBN, although in other cases the excited state molecule dissipates excess energy by *trans→cis* isomerisation about the N=N double bond. Thus azobenzene is not a practical photochemical source of phenyl radicals.

Radiolysis. Ionising radiation in the form of X- or γ-rays or α- or β-particles may interact with matter, and this affords another route to radical species. When an X- or γ-ray quantum interacts with a molecule, a fast electron is ejected. Molecular collision with β-particles, or with fast electrons derived from X- or γ-ray damage, may then wreak further havoc with many more molecules, usually by ejecting outer-shell electrons. The resulting 'secondary electrons' may themselves be sufficiently energetic to ionise still more molecules.

Initial loss of an electron from a neutral covalent molecule will result in a radical cation, but this may often react further, for example by deprotonation. One of the most abundant molecules in living organisms is water. Ejection of an electron from this gives the species H_2O^+, which, with a second molecule of water, forms H_3O^+ and a hydroxyl radical, HO·. Alternatively, recombination of H_2O^+ with an electron gives an excited-state water molecule which may dissociate into hydroxyl and a hydrogen atom. The further reactions of HO· play a major part in destroying tumour cells during radiotherapy, but they also damage healthy cells and contribute to the accompanying radiation sickness which is a commonly experienced side effect.

Molecule-induced homolysis. There is another circumstance in which undesirable and damaging hydroxyl radicals may arise in living organisms. It is as a result of the Fenton Reaction, in which iron(II) salts are oxidised by hydrogen peroxide in a simple bimolecular process (Eqn 2.15). This reaction is extremely rapid in aqueous medium. Fortunately, despite the importance of iron in living systems, and the formation of hydrogen peroxide by a number of perfectly normal biochemical processes, Fenton chemistry is not a problem in healthy tissues. This is because ferrous iron is strongly complexed in forms which are not easily oxidised. However, there are certain disease states associated with 'iron overload' in which excess iron may accumulate; Fenton chemistry ensues.

$$Fe^{++} \ + \ H_2O_2 \ \longrightarrow \ Fe^{+++} \ + \ HO^- \ + \ HO· \qquad (2.15)$$

As well as reacting with some transition-metal ions, a number of easily oxidised organic molecules will also react with various peroxides to form radicals. One instructive example is the room-temperature reaction of dibenzoyl peroxide with *N,N*-diphenylhydroxylamine. Provided that the hydroxylamine is in large excess, the reaction proceeds cleanly according to Scheme 2.6 (overleaf). It is not clear whether the first step (marked 'A'), i.e.

This discussion of the photolysis of di-*t*-butyl peroxide embraces what is sometimes referred to as 'The first law of photochemistry', which may be stated as follows: in order for light to bring about any chemical change it must first be absorbed.

the bimolecular reaction between peroxide and hydroxylamine derivative has the character of electron transfer, akin to the Fenton reaction, *followed* by proton transfer, or whether it is a one-step hydrogen-atom transfer, as implied in the scheme.

An interesting feature of this reaction sequence is the interception of the benzoyloxyl radical before it loses carbon dioxide. Decarboxylation of aromatic acyloxyl radicals is appreciably slower than that of their aliphatic counterparts.

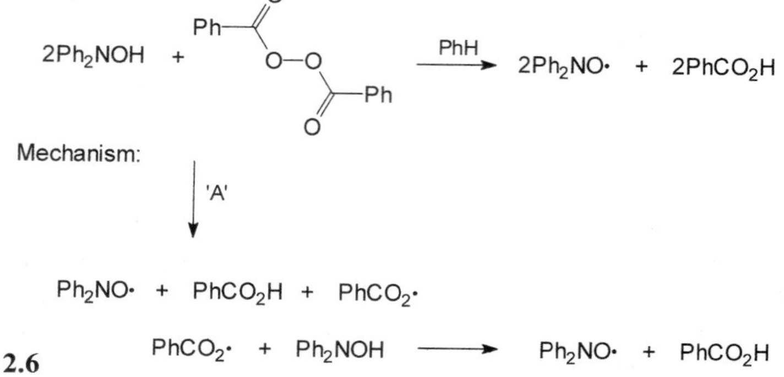

$$2Ph_2NOH + \text{(dibenzoyl peroxide)} \xrightarrow{PhH} 2Ph_2NO\cdot + 2PhCO_2H$$

Mechanism:

$$Ph_2NO\cdot + PhCO_2H + PhCO_2\cdot$$

$$PhCO_2\cdot + Ph_2NOH \longrightarrow Ph_2NO\cdot + PhCO_2H$$

Scheme 2.6

2.18

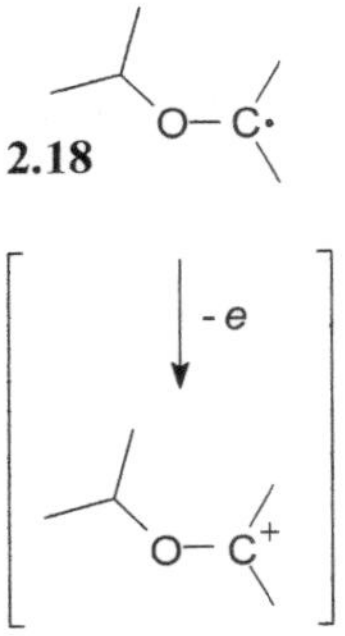

Radical-induced decomposition of peroxides, referred to earlier (p. 15), may parallel the above process. For example, at 80°C the half-life of dibenzoyl peroxide in dilute solution in benzene is about 4 hours, but in an aliphatic ether under similar conditions it is very much less. Radicals from thermolysis of the peroxide abstract hydrogen adjacent to ether oxygen. This results in an easily oxidised alkoxyalkyl radical (e.g. **2.18**) which may induce the decomposition of a second peroxide molecule.

Nevertheless, for cases such as that in Scheme 2.6, the designation 'molecule-induced homolysis' has been used, and there are other instances where the experimental evidence appears unequivocally to rule out the possibility of electron transfer. Perhaps the classic example of this is found in the spontaneous initiation of polymerisation of styrene. It is now known that this follows a very slow Diels-Alder dimerisation of styrene to give **2.19**; this then interacts with a further molecule of styrene by hydrogen transfer, restoring the aromatic ring to the dimer, but generating two (benzylic) radicals capable of diffusing apart and initiating polymerisation.

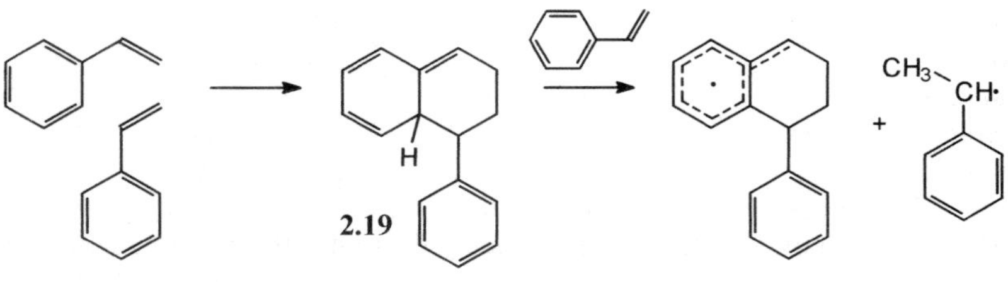

2.19

It should be noted that the doubly allylic hydrogen marked in structure **2.19** is held by an especially weak bond. When it is broken, the radical which is formed enjoys the aromatic stabilisation of a benzyl radical.

In recent years, many other reactions have come to light in which two hydrocarbon molecules will undergo this 'retrodisproportionation' (e.g. Eqn 2.16). In all cases, one of the reaction partners will have a particularly weak bond which, when broken, yields a highly resonance stabilised radical. Even so, many of the cases studied require elevated temperatures – the example shown occurs around 300°C – in contrast to the room-temperature initiation of styrene polymerisation.

(2.16)

Very recently, it has been suggested that the dehydrogenation of hydroaromatic compounds, for example dihydroanthracene (see Eqn 2.16) or tetralin (**2.20**), by quinones bearing electron withdrawing groups – the so-called 'high-potential' quinones, e.g. **2.21**–**2.23** – may be initiated by such a reaction (e.g. Eqn 2.17).

(2.17)

2.21

2.22

The cage effect. Homolytic rupture of a covalent bond, either by heat or by light, initially produces two radicals in close proximity. In solution, there will be a finite probability that the partners in this 'geminate pair'[*] will collide before they can diffuse apart. When it is remembered that radical-radical reactions commonly occur at the diffusion-controlled rate (simplistically, on every collision), it will be appreciated that a fraction of homolysis events does not result in freely diffusing radicals. Those pairs of radicals which recombine are said to have undergone 'geminate recombination' or to have recombined 'within the solvent cage'. If geminate recombination simply regenerates the starting material, it is not easily detected. But this is not always the case. The thermal decomposition of AIBN (**2.16**) in oxygen-free benzene nicely illustrates the phenomenon. This reaction occurs with the loss of nitrogen, and produces the dimer (**2.24**). Not surprisingly, if a substance which readily reacts with radicals (i.e. a 'radical scavenger') is added to the reaction mixture, the yield of **2.24** is reduced. Suitable scavengers might be styrene, or the stable radical DPPH (**1.2**). However, even large concentrations of scavengers fail to reduce the yield of dimer to zero. Experimental results are illustrated in Fig. 2.1, and

2.24

---

[*] From Gemini, the twins.

rationalised in Scheme 2.7, in which the concept of a solvent cage is represented by the bar placed over the initial radical pair.

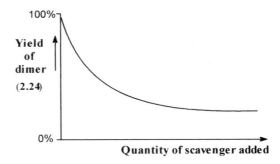

Fig. 2.1 Yield of dimer (**2.24**) from the thermolysis of AIBN in benzene as a function of the concentration of added radical scavenger.

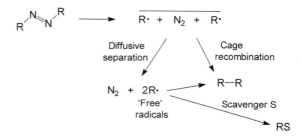

Scheme 2.7 Schematic representation of the cage effect in the decomposition of an aliphatic azo-compound. The initially formed pair of radicals may recombine before they diffuse apart. Only when they have separated to become 'free radicals' may we realistically begin to consider the competition between radical-radical and radical-molecule reactions.

It can be seen from this brief discussion how the adjective 'free' may nowadays be used to differentiate a 'diffusively free' radical from its 'caged' radical precursor.

Many subtleties of the cage effect are beyond the scope of an introductory discussion. For example, a distinction has been drawn between a primary radical pair, in which the initial partners remain adjacent, and a secondary pair, in which some solvent molecules intervene but for which there remains a significant probability of immediate diffusive re-encounter. Also important are experimental data on variation of the extent of cage recombination with pressure, and, more particularly, on its increase with increasing solvent viscosity. For more detailed discussion, the interested reader should consult the bibliography.

2.6   Redox chemistry of radicals

The one-electron oxidation or reduction of organic radicals may also be important in certain circumstances. For example, cyclohexadienyl radicals, which are intermediates in the important Birch reduction of benzene rings, undergo one-electron reduction in the course of the reaction (Scheme 2.8; see also discussion on p. 77).

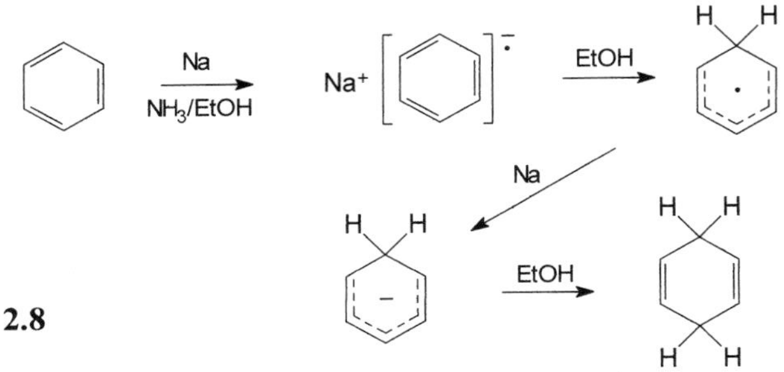

Scheme 2.8

One-electron transformations of radicals are particularly common in reactions catalysed by transition-metal ions. The formation of hydroxyl radicals by the Fenton Reaction in water saturated with benzene gives biphenyl and phenol. The proportion of phenol depends on the concentration of iron(III), which intercepts the intermediate hydroxycyclohexadienyl radical, as shown in Scheme 2.9 .

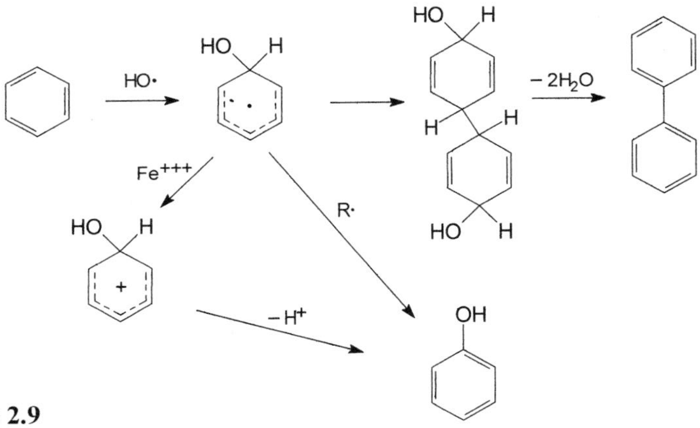

Scheme 2.9

This is a particularly complex area, not least because covalent bonding to transition metal ions may be a complicating factor. One notable example of this is found in cobalt chemistry. Whilst developments in this area have undoubtedly stemmed from efforts to understand the biological role of coenzyme B_{12} (adenosylcobalamin), they have led to some interesting and synthetically useful transformations. Thus reduction of cobalt(II) in its 'salophen' complex (**2.25**; abbreviated to Sal=Co(II)) gives a cobalt(I) product which, without isolation, reacts with organic halides to yield the alkylcobalt(III) complexes (**2.26**). These dissociate on heating or ultraviolet irradiation to give alkyl radicals. An interesting consequence in these systems is that the alkyl radicals are then oxidised to alkene, although various possibilities exist for intermolecular or intramolecular interception of the alkyl radical before oxidation occurs. Examples of this are illustrated in Scheme 2.10. The intramolecular case, which affords a β-lactam, is

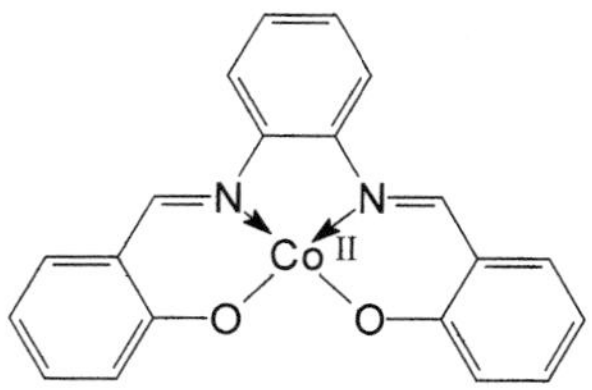

2.25 Cobalt(II) salophen ['Sal=Co(II)']

$\downarrow$ + e⁻

Sal=Co(I)⁻

$\downarrow$ RCH₂CH₂I

Sal=Co(III)⁺   I⁻

2.26 CH₂CH₂R

interesting in that it represents a rather unusual instance of a 4-*exo*-cyclisation.

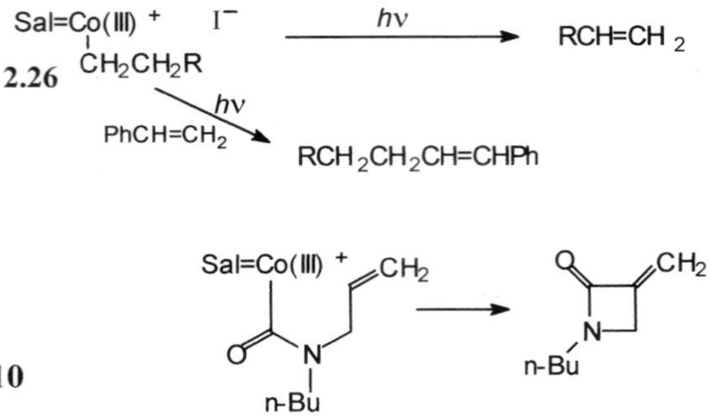

Scheme 2.10

The reader will probably be familiar with examples of the use of copper or copper ions in organic chemistry. A well-known example is the Sandmeyer reaction, in which aryl halides are made from diazonium salts. Radical chemistry is clearly involved here, but where copper salts mediate radical reactions, both electron-transfer processes and formation of organocopper species are known. The complexity of this area finds a dramatic illustration in the paint industry. We have noted previously that the 'drying' of oil-based paints is a radical-mediated autoxidation (see also Chapter 5). Commercial paints incorporate 'driers'. These are mixtures of organic-soluble carboxylate salts ('soaps') of redox-active metal ions, notably cobalt and manganese, which promote the autoxidation in a controlled fashion. But to this day, the precise formulation of these driers remains something of a 'black art'.

Radical ions. We shall take a closer look at one-electron oxidation or reduction of organic species in Chapter 6, but the example of the Birch reduction of benzene (Scheme 2.8) draws attention to the participation of radical ions in some organic reactions, in this case the benzene radical anion.

One interesting feature is also worth mentioning here. It is that when one-electron oxidation or reduction is operating in an organic reaction, it is often possible to conceive of more than one mechanistic pathway to the products – and the obvious one is not always correct. A good example is the electrochemical oxidation of acetate in the presence of naphthalene. Under apparent Kolbe conditions, the product is 1-acetoxynaphthalene, in which no decarboxylation of the acetoxyl radical has occurred. This was at first very puzzling, because in essentially all of its other reactions acetoxyl was known to undergo decarboxylation so rapidly that products retaining the elements of CO_2 were not observed. The explanation, however, is quite simple. Naphthalene is more easily oxidised by removal of an electron than is acetate, and the resulting naphthalene radical cation is then trapped by acetate (Scheme 2.11).

It is also pertinent to mention here that, unlike the great majority of reactions of electrically neutral radicals, redox reactions and electrochemical one-electron redox potentials, as well as reactions of radical ions, are likely to show pronounced solvent dependence.

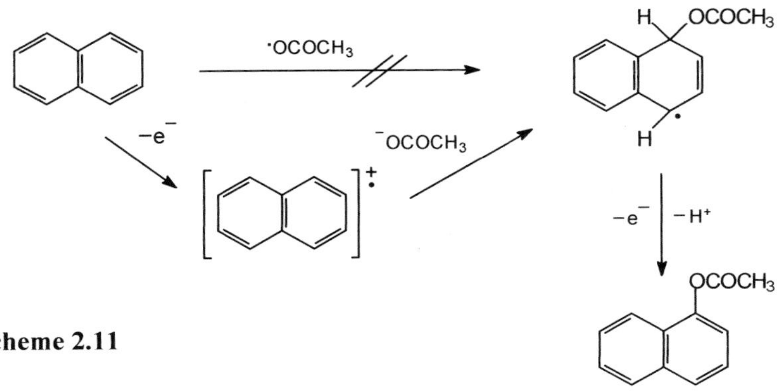

Scheme 2.11

2.8   Radical reactions exemplified

Classical examples of radical reactions, such as simple atom substitutions and simple additions have already been encountered, the latter being exemplified both by polymerisation and by formation of 1:1 adducts. In this section we shall introduce a few more examples of these and other preparatively useful reactions that involve the various unit steps described in earlier sections of this Chapter. However, examination of other, arguably more interesting examples will be deferred until we have established some understanding of the factors which determine patterns of reactivity in radical chemistry. These will be addressed in Chapter 3.

Nitrosation of cyclohexane. A commercially important reaction is the Japanese Toray process, one of the methods of transforming cyclohexane into caprolactam, which is a precursor of the polyamide 'Nylon-6'. This depends upon the photolysis of nitrosyl chloride in cyclohexane, and proceeds as in Scheme 2.12. The lack of selectivity in chlorination reactions has already been commented upon, but, by virtue of symmetry, cyclohexane can form only one monosubstitution product. The surprising feature of this process is the extraordinary efficiency with which cyclohexyl radicals react with NO. One possible interpretation of this might have been that there is a chain

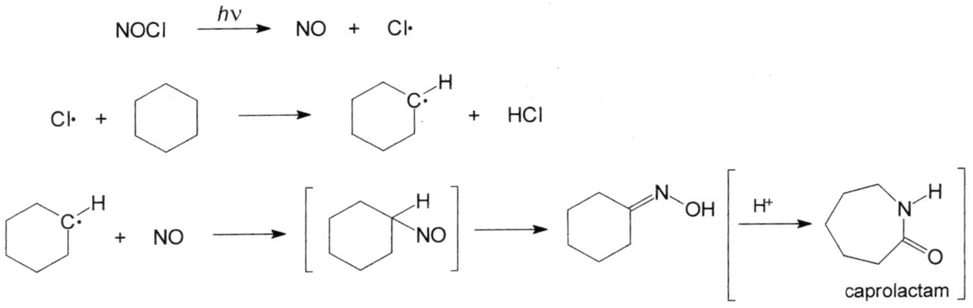

Scheme 2.12

The reader may like to ponder why it is in Scheme 2.12 that cross-coupling of NO and cyclohexyl should be so successful in competition with the diffusion-controlled self reaction of two cyclohexyl radicals (the explanation is discussed in Chapter 3).

N.B. Quantum Yield = (No. of molecules undergoing change)/ (No. of light quanta absorbed).

reaction, in which cyclohexyl radicals react with nitrosyl chloride, displacing a chlorine atom that propagates the chain. But this would be a photocatalytic process with a quantum yield $\gg 1$. In fact the reaction has a quantum yield of unity, fully consistent with Scheme 2.12.

Like the Kolbe and Teuber reactions, this is an example of a preparatively useful radical reaction which does *not* have a chain mechanism.

Allylic bromination. Another long-established substitution reaction involves allylic bromination using *N*-bromosuccinimide (NBS). This *is* a chain reaction, initiated by peroxide, the most familiar illustration of which, shown in Eqn 2.17, conceals several important features. Crucial to the success

$$\qquad\qquad\qquad\qquad\qquad\qquad\qquad (2.17)$$

2.27

of the reaction is the relative weakness of the allylic C–H bond. This is a consequence of resonance stabilisation in the derived allylic radical (**2.27**). Now bromine may attach to this radical at one of two sites. Whilst symmetry simplifies matters in the case of cyclohexene, and only one product is possible, it is clear that had the substrate been, say, 4-methyl-cyclohexene, a complex array of mono-bromo compounds would be possible (Scheme 2.13).

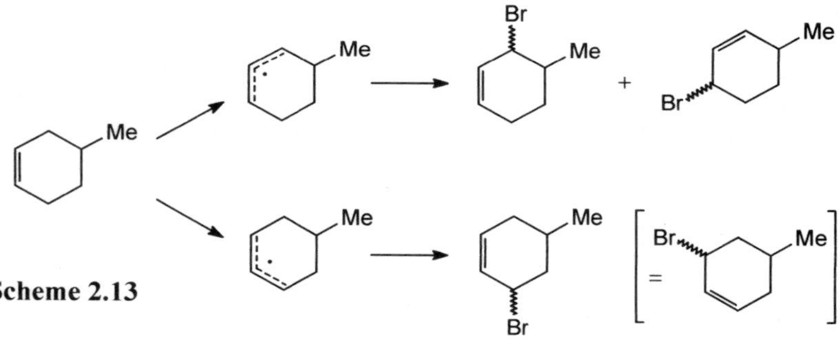

Scheme 2.13

A more subtle complication is encountered when the chain mechanism for allylic bromination is analysed. Detailed study of the reaction has revealed that, under normal circumstances, the role of NBS is to maintain a low concentration of molecular bromine in the solvent (CCl$_4$, in which both NBS and the succinimide by-product are only sparingly soluble). The chain reaction then involves hydrogen abstraction by bromine. This is quite different from the radical-initiated reaction of higher concentrations of bromine with cyclohexene, which results in vicinal addition. The contrasting sets of behaviour are a consequence of the ready reversibility of bromine-atom addition to many alkenes. If only a low concentration of molecular bromine is present, interception of the bromine-atom adduct (**2.28**) does not compete with the irreversible allylic substitution. The low concentration of bromine is maintained by reaction of NBS with HBr. The mechanism is set out in Scheme 2.14.

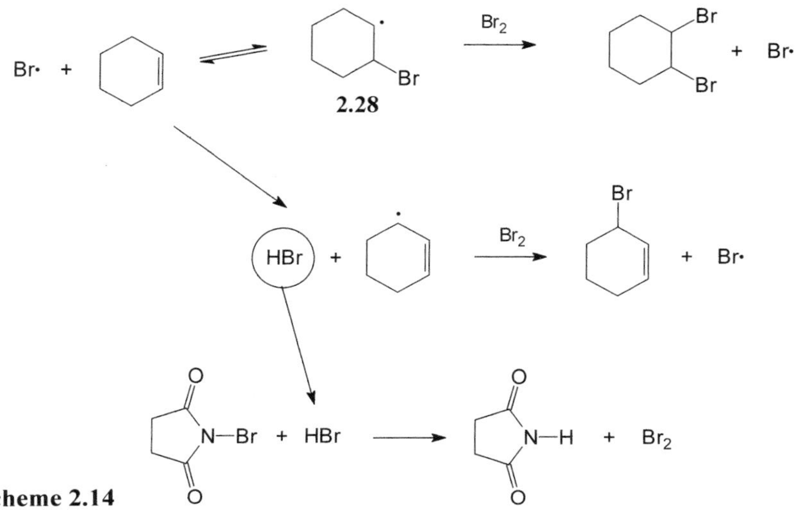

Scheme 2.14

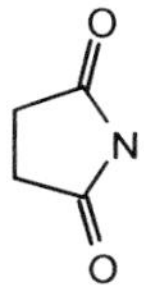

If the NBS reaction is carried out in the presence of an *alkene lacking allylic hydrogen atoms*, for example 3,3-dimethyl-1-butene, the bromine is removed by addition to the alkene double bond; bromine substitution now takes place with a dramatically different selectivity, since hydrogen abstraction is by the succinimidyl radical (**2.29**), and the resulting hydrocarbon radical abstracts bromine from NBS. Succinimidyl is a much more reactive species than is a bromine atom. Remembering the difference in selectivity exhibited by chlorine atoms and bromine atoms, introduced in Chapter 1, the behaviour of the succinimidyl radical is more akin to that of $Cl\cdot$ than to that of $Br\cdot$.

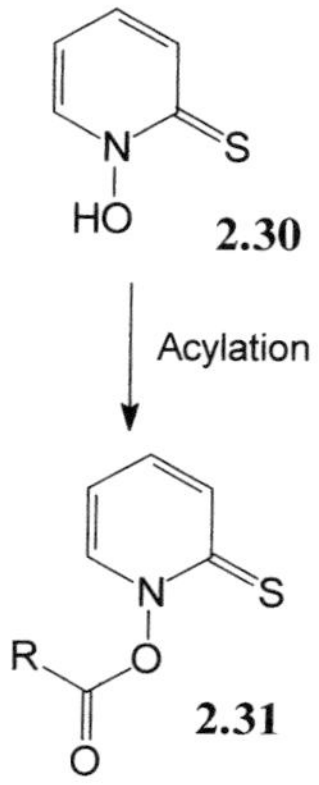

Functional-group transformations using derivatives of N-hydroxypyridine-2-thione. One special case which, like tin hydride chemistry, has found a prominent position in the armoury of the synthetic chemist, depends on the susceptibility of *O*-acyl derivatives (**2.31**) of *N*-hydroxypyridine-2-thione (**2.30**) to radical addition at the sulphur atom. Photolysis of **2.31** in a relatively unreactive solvent such as benzene initiates a chain reaction, the overall result of which is a decarboxylative rearrangement. The mechanism of this is indicated in Scheme 2.15. Critically, fragmentation of the radical formed by addition at sulphur (a 'pyridinyl' radical, **2.32**) is succeeded by decarboxylation of the resulting acyloxyl radical, and the cycle is repeated.

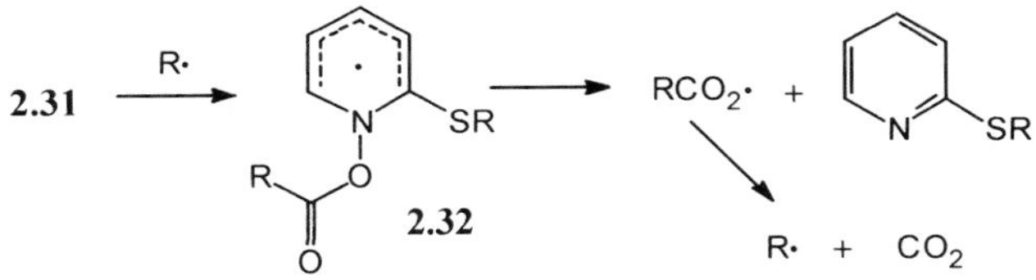

2.29

2.30

Acylation

2.31

Scheme 2.15

Of itself, little importance might attach to such a process, but it was appreciated that the chain might be varied by intercepting the intermediate R· radical. Since the thiones (**2.31**) are readily available from carboxylic acids, this has led to a versatile set of functional group interconversions of the kind RCO₂H → RX, often in excellent yield. Some of these are indicated in Table 2.2. The essential feature of these transformations, over and above what is outlined in Scheme 2.15, is the addition of the two steps generalised in Eqns 2.18 and 2.19. For optimum results, the concentrations of the reagents must be carefully adjusted to minimise the attack of R· on **2.31**, but under favourable circumstances preparative yields well in excess of 70% have been achieved.

$$R\cdot \;+\; XY \longrightarrow RX \;+\; Y\cdot \tag{2.18}$$

$$\tag{2.19}$$

The acyloxypyridinethiones are now frequently referred to as 'PTOC esters'.

Table 2.2 A selection of products available from *N*-acyloxypyridine-2-thiones (**2.31**)

Reagent (X–Y)	Product (R–X)	Y in **2.33**
CCl₄	RCl	CCl₃
BrCCl₃	RBr	CCl₃
PhSSPh	RSPh	SPh
PhSeSePh	RSePh	SePh
HSnBu₃	RH	SnBu₃
CH₂=CHCN	RCH₂CH₂CN	SnBu₃
+ HSnBu₃		

In several respects, these transformations may have distinct advantages over the more conventional nucleophilic aliphatic substitution routes to RX. The alkyl group, R, may be primary, secondary, or tertiary, and, in the last case, this includes the possibility of reactions at a bridgehead position, permitting *inter alia* the efficient production of a variety of 1-substituted adamantane derivatives (e.g. **2.34**). Furthermore, groups such as hydroxyl are unlikely to interfere. For making halogen compound from carboxylic acids, the reaction is far more versatile than is the classical Hunsdiecker reaction of silver carboxylate with elemental halogen. Furthermore, variants which have been developed are capable of transforming RCO₂H into ROOH, ROH, RPO₃H₂, etc.

2.34

Aromatic arylation. A very different example in which the presence of hydroxyl is not a complicating factor is found in a case of aromatic substitution. Ultraviolet photolysis of 3-iodophenol in benzene gives a good yield of 3-hydroxybiphenyl. The reaction is depicted in Scheme 2.16, light energy here being sufficient to break the relatively strong aryl–iodine bond.

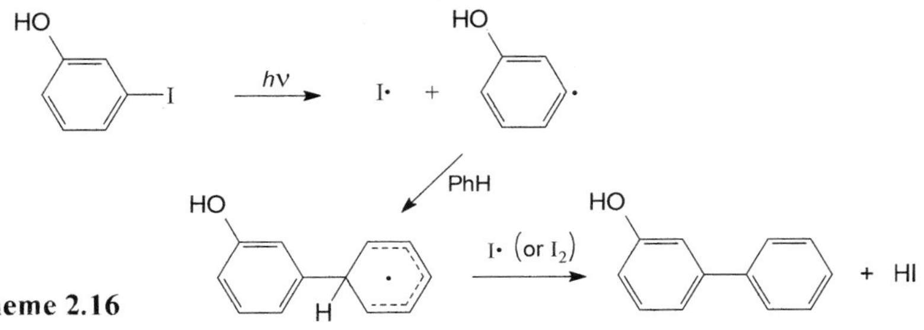

Scheme 2.16

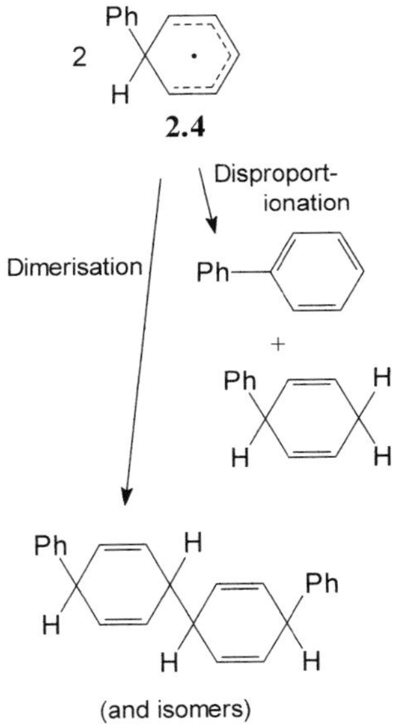

The formation of simple biaryls by radical arylation is most successful when the substrate is symmetrical (e.g. benzene itself, or a symmetrically *para*-disubstituted benzene), in which case only one substitution product is possible (*cf.* Chapter 1). It is also most efficient when there is a good oxidising species present – in the present example, this is iodine. When this is absent, as is the case with some other aryl radical sources, biaryl yields may be greatly reduced because the intermediate arylcyclohexadienyl radicals (e.g. **2.4**) may dimerise or disproportionate (*cf.* hydroxylation example on p. 21).

Aryl iodide photochemistry, like photonitrosation, is a synthetically useful non-chain process. Mechanistically quite different, another valuable radical reaction which does not have a chain mechanism provides a surprisingly efficient synthesis of cyclodecane. This medium-ring alicycle is notoriously difficult to prepare by classical cyclisation procedures because of the pronounced non-bonded interactions which develop as the ends of a ten-carbon chain are brought together, and which, in the product, are designated 'transannular' strain. The synthesis depends on pyrolysis of the bis-peroxide **2.35**. This is readily available from hydrogen peroxide and cyclohexanone. The reaction is depicted in Scheme 2.17, in which peroxide homolysis, radical fragmentation, and *intramolecular* radical coupling lead to the desired product in reasonable yield (*c.* 50%).

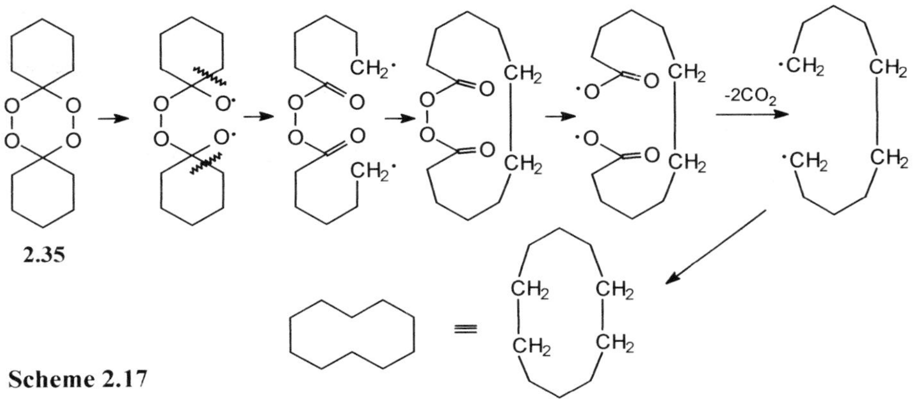

2.35

Scheme 2.17

Cysteine

Cystine

Thiol oxidation. It will come as no surprise that the various oxidation states of sulphur which are found in stable compounds incorporating the element are paralleled by a rich variety of radical chemistry. One-electron oxidation of dialkyl sulphides (R_2S), for example, can form dimeric radical cations ($R_2SSR_2^{\cdot+}$), and there is an extensive chemistry of sulphonyl ($RSO_2\cdot$) and sulphinyl ($RSO\cdot$) radicals. However, of especial importance is the facile oxidation of the weak S–H bond in thiols to give disulphides. The reverse process, the reduction of disulphides to thiols is also well known, and a variety of reagents is available which will bring about these changes in the laboratory. The reaction has special significance in biological systems where the interconversion of cysteine and cystine residues in peptides and proteins is of such great importance. A familiar example of this is found in the alignment of protein molecules in hair. These are held together by disulphide links. The result of chemically breaking these and reforming them in different locations after the hair has been curled or straightened is the 'permanent wave'. Far more important, however, is the involvement of this chemistry in the mode of action of various redox enzymes. For example, all mammalian cells contain peroxidase enzymes which facilitate reduction of hydrogen peroxide to water; these depend on the presence of the thiol 'glutathione' (**2.36**), a tripeptide, as hydrogen donor. The glutathione is regenerated from the corresponding disulphide by a further enzyme-catalysed reaction.

2.36

The ease of thiol oxidation is reflected in certain synthetic strategies in which thiols have been added to reaction systems in order to intercept a variety of intermediate radicals. A good example is found in a hydroperoxide synthesis using the pyridinethione chemistry discussed earlier. To achieve this the system contains *t*-butyl mercaptan (*t*-BuSH) and is oxygenated. The alkyl radical is trapped by O_2 to give ROO·, which is in turn reduced by thiol to hydroperoxide. The *t*-butylthiyl radical continues the chain by adding to the thione sulphur. Selenols, e.g. PhSeH, are even more effective reducing agents for alkyl radicals, with which they react at rates close to the diffusion-controlled limit. They present handling problems, however, both because of the stench and because of rapid air oxidation to diselenides. Frequently these problems can largely be circumvented by reducing a (much less offensive) diselenide, e.g. PhSeSePh, to selenol *in situ*, using sodium borohydride.

3 Energetics, kinetics, and mechanism

3.1 Introduction

In order properly to grasp the potential of radical chemistry, it is necessary to understand some basic aspects of the energetics and kinetics of radical processes and the factors which determine them. Fortunately, there are some simplifying considerations, such as the fact that reaction rates are commonly unaffected by solvent polarity, and that, for closely related reaction steps, there is a simple relationship between rate of reaction and heat of reaction.

Nevertheless, at first sight many features may seem puzzling. For example, how can it be that a chain propagating step, for which a representative rate constant may be of the order of 10^2-10^5 M^{-1} sec^{-1}, can compete successfully with a termination process whose rate constant is close to 10^{10} M^{-1} sec^{-1}? To examine this, we shall begin with a somewhat over-simplified analysis of the decomposition of a dilute solution of a diacyl peroxide [$(RCO_2)_2$] in an unreactive, non-polar medium such as benzene. We shall assume that the reactions shown in Scheme 3.1 provide a complete picture of the reaction steps involved. The initial concentration of the peroxide will be taken as 0.005 M, and we shall suppose that the half-life for decomposition of the peroxide is c. 4 hours at 80°. This corresponds to a first-order rate constant, $k_1 = c$. 10^{-4} sec^{-1}. For radical destruction, $2k_2 = c$. 10^{10} M^{-1} sec^{-1}. Since the decarboxylation of the acyloxy radicals is very fast, and may be regarded as kinetically unimportant, we may write:

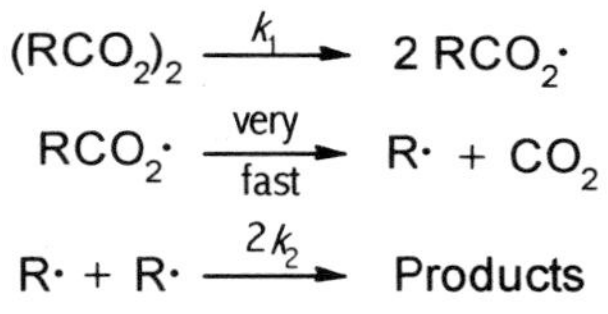

$$d[R\cdot]/dt = 2k_1[(RCO_2)_2] - 2k_2[R\cdot]^2$$

Furthermore, since $[R\cdot]$ must be very small, it is reasonable to apply the stationary state approximation by setting $d[R\cdot]/dt$ = zero. Hence:

$$2k_1[(RCO_2)_2] = 2k_2[R\cdot]^2$$

and therefore:

$$[R\cdot] = (2k_1[(RCO_2)_2]/2k_2)^{\frac{1}{2}}$$

Inserting the values which we have chosen for the parameters on the right-hand side of this equation, we find that at the outset of the reaction, when a negligible amount of peroxide has decomposed, $[R\cdot] = (2 \times 10^{-4} \times 0.005 / 10^{10})^{\frac{1}{2}}$, $i.e.$ $[R\cdot] = 10^{-8}$ M.

$(RCO_2)_2 \xrightarrow{k_1} 2\,RCO_2\cdot$

$RCO_2\cdot \xrightarrow[\text{fast}]{\text{very}} R\cdot + CO_2$

$R\cdot + R\cdot \xrightarrow{2k_2}$ Products

Scheme 3.1

This is a very important result because it gives a realistic idea of the likely concentration of reactive radicals in a system in which they are being generated by a conventional initiator in a mobile solvent, provided that their self-reaction occurs at the diffusion-controlled rate. With much higher initiation rates, e.g. with very high concentrations of initiator at a temperature where its half-life is greatly reduced, or under conditions of intense ultraviolet irradiation, stationary concentrations of radicals may rise to 10^{-6} M or higher. On the other hand, in biological systems, where initiation rates are much slower, stationary-state concentrations may be significantly below 10^{-8} M. What is clear, however, is that, for direct detection to be achieved, highly sensitive methods must be found.

Let us now extend our kinetic analysis a little further, and consider the use of our peroxide to initiate a chain reaction in which an alkene is polymerised. Once again we shall make an oversimplifying assumption. This is that when R· reacts with one molecule of alkene we get a new radical, R(1)·, which is kinetically indistinguishible from R·, i.e. it reacts either with itself or with R· at the diffusion-controlled rate, and it reacts with a further molecule of alkene at the same rate as does R·. Let us also suppose that the rate constant for the propagation (i.e. addition) reaction, k_p, is 10^4 M^{-1} sec^{-1}, and, furthermore, that R(2)·, formed by addition of R(1)· to alkene, behaves similarly, and so on. Then we can generalise as follows (where R(n)· includes R·):

It should perhaps be emphasised that, in the context of polymerisation, the adjective 'chain' may be used both in the kinetic sense of a chain of repeating events, and in the context of a 'long-chain' polymer.

$$\text{R(n)·} \quad + \quad \text{Alkene} \quad \xrightarrow{\;k_p\;} \quad \text{R(n+1)·} \tag{3.1}$$

and:

It is customary to designate the rate constant for termination as $2k_t$. The multiplier 2 is included since two radicals are destroyed.

$$\text{R(n)·} \quad + \quad \text{R(n)·} \quad \xrightarrow{\;2k_t\;} \quad \text{non-radical products} \tag{3.2}$$

If we take the initial peroxide concentration to be the same as that in the reaction with no alkene present, and assume an initial alkene concentration of 1 M, then at the beginning of the reaction, with negligible consumption of reagents, it should be obvious that $\sum[\text{R(n)·}]$ will be 10^{-8} M. This is because, for the stationary state, radical production from the initiator and radical removal in the termination step must balance; initiation is identical to the situation with no alkene added, and termination is also proceeding at the same rate, except that R· is replaced by R(n)·.

We are now equipped to make an important comparison. The initial rate of consumption of R(n)·, i.e. d[R(n)·]/dt, in the propagating step, Eqn 3.1, will be $10^4 \times 10^{-8} \times 1 = 10^{-4}$ M sec^{-1}, whilst that in the terminating step, Eqn 3.2, will be $10^{10} \times 10^{-8} \times 10^{-8} = 10^{-6}$ M sec^{-1}. It is therefore 100 ($= 10^{-4} / 10^{-6}$) times *more likely* that the propagating radical will be consumed by reaction with alkene than will be destroyed in a termination step. Alternatively, we may say that the 'kinetic chain length' is 100.

Among the oversimplifying assumptions in this treatment is that, in practice, as the polymer chain grows, there is a perceptible diminution in the value of both k_p and $2k_t$. This effect is accentuated by a gradual increase in viscosity.

The object of this simplified analysis, which admittedly disregards several complicating features of a real polymerisation reaction, but uses 'reasonable' rate constants, has been to demonstrate how it is that 'slow' propagation reactions may compete very successfully with diffusion-controlled termination. This is, of course, the essence of all free-radical chain reactions.

In the preceding two chapters, reactivity has occasionally been related, in a qualitative way, to either the strength of a bond being broken, or to stabilisation in the reacting radical. For example, it was seen that the phenyl radical, in which the unpaired electron is localised in an *sp²* orbital, is sufficiently reactive to attack benzene, whereas the resonance-stabilised triphenylmethyl clearly is not. Similarly, benzene has been referred to as a relatively unreactive solvent. This contrasts with, for example, toluene, where many radical species are sufficiently reactive to remove the benzylic hydrogen – which is less strongly bound than are those on the benzene ring. In order to explore the factors which influence the actual rates of radical reactions, it is instructive first to examine some bond-energy data, and relate this to the structures and stabilities of various representative radicals.

3.2   Radical structure and stability

It will be appreciated that species which are seldom obtained in solution at concentrations greater than micromolar have been slow to yield to quantitative experimental study. Not surprisingly, therefore, revisions of quite fundamental data on, for example, the C–H bond dissociation energy (b.d.e.) of ethane, or the absolute rate constants for reactions of ethyl radicals, have appeared in quite recent literature. Indeed, for radical chemistry, modern molecular orbital calculations, at the highest levels of theory, now compare in reliability with experimental methods for the determination of radical energetics. In many cases they also come close to reproducing with acceptable accuracy the activation barriers for radical-molecule reactions.

Carbon-Centred Radicals. The simplest carbon-centred radical is methyl. This is known from spectroscopic studies to be planar, with the unpaired electron localised in a *p*-orbital (**3.1**). (Not surprisingly, pyramidal distortion is significantly easier than it is for a methyl cation, as a result of the repulsion between the single electron and C–H bond-pair electrons). The C–H bond-dissociation energy of methane is *c.* 435 kJ mol^{-1}. *sp³*-Hybridised C–H bonds in other hydrocarbons are weaker than this, principally because of stabilising effects in the derived radical. Some data are given in Table 3.1, overleaf.

3.1   **3.2** Benzyl   **3.3** Allyl

N.B. No attempt has been made in formulae **3.2** and **3.3** to represent the *p*-orbital coefficients in the π-SOMO. Indeed, in the SOMO of the allyl radical, there is a node at the central carbon atom.

The resonance stabilisation of benzylic and allylic radicals has been mentioned previously. In both cases, the SOMO (p.6) embraces the whole π-system. It is important to realise that the π-interaction restricts rotation about the phenyl–methylene bond in benzyl (**3.2**), and that the allyl structure in its ground state is likewise coplanar (**3.3**). Conversely, should allylic hydrogen be forced into a position orthogonal to the adjacent π-electrons, as in a

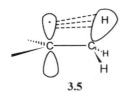

3.4

bridged alkene (e.g. **3.4**), there will be no significant weakening of the C–H bond since electron overlap in the derived radical is precluded.

Table 3.1 Dissociation energies of bonds* between hydrogen and sp^3-hybridised carbon in representative Hydrocarbons.

	Hydrocarbon	C–H b.d.e. (kJ mol^{-1})
1	CH$_3$–H	439
2	CH$_3$CH$_2$–H	422
3	(CH$_3$)$_2$CH–H	412
4	(CH$_3$)$_3$C–H	402
5	CH$_2$=CHCH$_2$–H	362
6	(CH$_2$=CH)$_2$CH–H	347
7	C$_6$H$_5$CH$_2$–H	370

*Strictly, in both this and subsequent Tables, the figures represent gas-phase enthalpies of bond homolysis.

3.5

The modest stabilisation found with alkyl-substituted methyl radicals (entries 2-4 in Table 1) can be attributed in part to hyperconjugation (**3.5**). This type of interaction of the unpaired electron with adjacent, properly orientated σ-bonding electrons is clearly revealed both by theory and by spectroscopic data (see Chapter 4, p. 57). But there is a second reason why alkyl substitution contributes to C–H bond weakening. This is the release of steric compression as the alkyl substituents move apart to form a planar radical (Fig. 3.1).

Somewhat unexpectedly, it has been convincingly demonstrated that *t*-butyl, unlike methyl, does not, in fact, have a planar energy minimum, as implied in Fig. 3.1. Rather, it is slightly pyramidalised, with a tiny (*c*. 2kJ mol^{-1}) barrier to inversion.

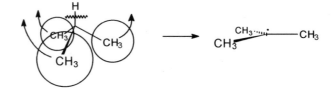

Fig. 3.1 Diagram showing the methyl groups of 2-methylpropane moving apart as the C–H bond breaks to leave a *t*-butyl radical.

Not surprisingly, when the hydrocarbons listed in Table 3.1 are compared with respect to their reactivity towards hydrogen abstraction by a specific radical, e.g. a bromine atom, those with the weakest C–H bonds react most rapidly. For the reaction with bromine, toluene is of the order of 5000 times more reactive than is ethane at 40°C. Conversely, when the radicals derived from these hydrocarbons react with a given substrate, those that form the strongest bonds to hydrogen tend to react most rapidly. Thus methyl abstracts a chlorine atom from CCl_4 some 1000 times more rapidly (at 60°C) than does benzyl.

Despite these reactivity differences, all of the radicals derived from hydrocarbons listed in the Table undergo bimolecular *self-reaction* (dimerisation or disproportionation) at essentially the same (i.e. the diffusion-controlled) rate. For radicals having termination reactions that are significantly slower than the diffusion-limited rate, the designation 'persistent' has been adopted. By virtue of their lower rates of self reaction, persistent radicals may accumulate in solution to concentrations significantly higher than micromolar, thus lending themselves more easily to direct spectroscopic examination. The stable radicals referred to in Chapter 1, as well as triphenylmethyl which, in solution, is in equilibrium with its dimer, might all be designated persistent. It is important, however, to recognise that persistence and stabilisation are quite distinct phenomena. Stabilisation is a thermodynamic property, normally arising from electron delocalisation. Some radicals which are persistent may enjoy very little stabilisation. A good example is the tertiary alkyl radical, triisopropylmethyl (**3.6**), which adopts the conformation shown. This radical is insufficiently persistent to be isolated, but, because of strong steric interference at the radical centre, dimerisation is inhibited. Similarly, the isopropyl groups must twist into a very strained geometry to permit disproportionation to give an alkene. The result is a rate constant for self-reaction of only 10^3 M^{-1} sec^{-1} at 25°C.

The emphasis in this section has been upon radicals, like methyl, in which the unpaired electron resides in a *p*- or delocalised π-orbital, so that the atom(s) at the radical centre(s) lie in the nodal plane of the SOMO (e.g. **3.1** **3.2** and **3.3**). Such radicals are commonly referred to as 'π-radicals'. The designation 'σ-radical' is used for those species in which the unpaired electron occupies an orbital with appreciable *s*-character, and which therefore have some unpaired electron density at the nucleus. As is the case with phenyl, mentioned earlier, such radicals commonly form rather strong bonds to hydrogen (Table 3.2).

3.6

An unstabilised silacyclobutyl radical has recently been isolated:

Table 3.2 C–H bond dissociation energies of molecules which produce carbon-centred σ-radicals

Hydrocarbon	C–H b.d.e. (kJ mol^{-1})
$CH_2=CH-H$	465
$CH\equiv C-H$	556
C_6H_5-H	465
CF_3-H*	446

*The trifluoromethyl radical is pyramidal, with the unpaired electron in an sp^3-like orbital. Appreciably pyramidalised structures are found for other radicals having electronegative substituents, e.g. $(RO)_2CH\cdot$.

Non-carbon-centred radicals. The high reactivity of a hydroxyl radical, noted earlier (p. 5), may be associated with the great strength of the O–H bond in the water molecule. The effects of substituents on oxygen are rather

A consequence of the strengths of bonds to hydrogen in compounds listed in Table 3.2 is the extremely low reactivity which they exhibit towards hydrogen abstraction. Of course, radical addition is possible in the first three cases, although a very reactive radical is required to disrupt the π-electron system of benzene. One other hydrocarbon which is particularly resistant to radical attack is cyclopropane. For this, it will be remembered that the hybrid orbital on carbon which is involved in bonding to hydrogen approximates to sp^2 character. The C–H b.d.e. is *c.* 445 kJ mol^{-1}.

Acyl radicals are also σ-radicals, but the strengths of aldehyde C–H bonds are much lower than those in Table 3.2 (b.d.e. $CH_3C(=O)-H$ *c.* 365 kJ mol^{-1}). In an acyl radical there is stabilising interaction between the single electron and a lone-pair on oxygen ($\leftrightarrow RC^-O\cdot^+$).

greater than is the case with carbon, as indicated by the homolytic bond-dissociation energy data for methanol and phenol in Table 3.3. Particularly interesting are the cases of hydroperoxides (including hydrogen peroxide) and hydroxylamines, where the corresponding peroxyl radicals, and more especially the nitroxides, depicted in structures **3.7** and **3.8** respectively, experience marked stabilisation from dipolar resonance contributions.

$$R-O-O\cdot \longleftrightarrow R-\overset{\cdot\,+}{O}-O^- \qquad \overset{R}{\underset{R}{N}}-O\cdot \longleftrightarrow \overset{R}{\underset{R}{\overset{\cdot\,+}{N}}}-O^- \qquad H-O-H \longleftrightarrow H-O^-\,H^+$$

$$\textbf{3.7} \qquad\qquad\qquad \textbf{3.8} \qquad\qquad\qquad \textbf{3.9}$$

The greater strength of O–H in water compared with C–H in methane can in turn be associated with a strong dipolar contribution to the σ-bond, (**3.9**). This is of even greater importance in H–F (b.d.e. = 570 kJ mol^{-1}).

Bonds to hydrogen from second-row elements and beyond are generally weaker than those for their first-row counterparts, and the interaction with substituents is diminished, *cf.* MeS–H and PhS–H in Table 3.3.

Table 3.3 Dissociation energies of bonds between hydrogen and a heteroatom

X–H	b.d.e. (kJ mol^{-1})	X–H	b.d.e. (kJ mol^{-1})
Cl–H	431	CH$_3$S–H	366
Br–H	366	C$_6$H$_5$S–H	330
I–H	297	HOO–H	368
HO–H	498	(*t*-Bu)$_2$NO–H	300
CH$_3$O–H	440	(CH$_3$)$_3$Si–H	388
C$_6$H$_5$O–H	356	(CH$_3$)$_3$Sn–H	310

3.3 The effects of bond strength and various other factors upon reaction rate

Atom-transfer reactions. Hydrogen-atom abstraction from methane by a chlorine atom is particularly instructive. This is a very fast reaction, with a rate constant of approximately 10^7 M^{-1} sec^{-1} at 25°C and an activation barrier of only about 15 kJ mol^{-1}. Inspection of the available data (Tables 3.1 and 3.3) shows that the reaction is almost thermoneutral. On the left-hand side of

$$CH_3-H \ + \ Cl\cdot \ \longrightarrow \ CH_3\cdot \ + \ H-Cl \qquad (3.3)$$

Eqn 3.3 the C–H bond has a bond dissociation energy of *c.* 439 kJ mol^{-1}. This is replaced, on the right hand side, by an H–Cl bond which has a dissociation energy of 431 kJ mol^{-1}. From these data, the reaction is weakly endothermic, and the transition state is only *c.* 7 kJ mol^{-1} above the energy of the products. This is depicted in Fig. 3.2. What is clear, therefore, is that, as the reaction path is traversed, there is an efficient bond reorganisation *via* the three-centre transition state (**3.10**), with an overall energy variation which is astonishingly small.

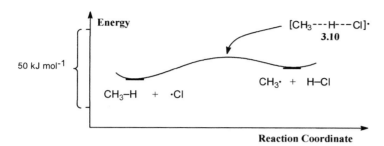

Fig. 3.2 Diagram showing the variation of energy with progress along the reaction coordinate for abstraction of a hydrogen atom from methane by a chlorine atom.

Since the primary, secondary and tertiary C–H bonds of alkanes are weaker than that in methane (Table 3.1), we might reasonably expect that other aliphatic hydrocarbons would react with chlorine atoms even faster. On the other hand, reaction with bromine atoms will be endothermic, and consequently much slower. We learnt in Chapter 1 that, whilst alkane chlorination seems almost random, bromination is remarkably selective. It turns out that by simply skewing the graph of Fig. 3.2 to the heats of reaction for hydrogen abstraction by Cl· or Br· we can arrive at a semi-quantitative picture of the experimental selectivities found in alkane halogenation by these two elements. Consider firstly the chlorination of 2-methylpropane, $HC(CH_3)_3$. Both possible hydrogen abstraction steps will be exothermic. For abstraction of hydrogen from a methyl group the reaction replaces a primary C–H bond having a b.d.e. $= 422$ kJ mol^{-1} with an H–Cl bond (b.d.e. $= 431$ kJ mol^{-1}). The reaction is therefore exothermic by 9 kJ mol^{-1}. For the tertiary position, the corresponding value is *c.* 29 kJ mol^{-1}. By the appropriate skewing of Fig. 3.2 for each process, we arrive at the two curves shown in Fig. 3.3. It is clear that both activation barriers will be very small,

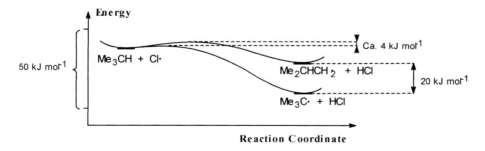

Fig. 3.3 Reaction coordinate diagram for the Cl· / 2-methylpropane reaction.

as will the difference between them. An estimate for this difference from the graphs might be *c*. 4 kJ mol^{-1}. If both reactions obey the Arrhenius equation, and have the same pre-exponential factor, this corresponds, at 25°C, to a rate difference of *c*. 5 in favour of the tertiary position. Remembering that there is a statistical factor of 9 (the number of primary hydrogens), we might therefore predict that the ratio of primary chloride, $(CH_3)_2CHCH_2Cl$, to tertiary chloride, $(CH_3)_3CCl$, would be approximately 9/5:1, or 64:36. The experimental result is about 60:40, which shows remarkably close agreement for such a simplistic approach.

Extending this procedure to an examination of bromination of 2-methyl-propane produces Fig. 3.4. Both reactions are now endothermic (b.d.e. for H–Br = 366 kJ mol^{-1}), and the graphs are therefore skewed in the opposite sense to before. In this case, the difference in activation barriers appears to be nearly as great as is the difference between the endothermicities of the two reactions, say 18, compared with *c*. 20 kJ mol^{-1}. Allowing once more for the statistical factor of 9 in favour of reaction at the primary position, this still translates into a 99% preference for reaction at the tertiary site, also in good agreement with experiment.

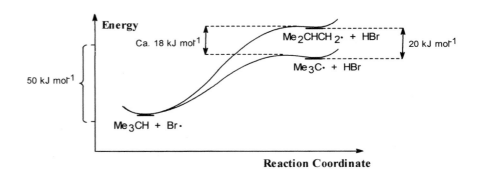

Fig. 3.4 Reaction coordinate diagram for the Br· / 2-methylpropane reaction.

One assumption which is implicit in these arguments is that selectivity in hydrogen abstraction by a halogen atom translates directly into the proportion of products formed. Provided that there is no exchange between alkyl radical and alkane (a possibility which will be excluded later), and provided that there is a moderate concentration of molecular halogen present, with which reaction of alkyl radicals is highly exothermic and essentially diffusion controlled, this assumption is valid. (Note that although Fig. 3.4 suggests that the rate of the reverse reaction, i.e. that between the alkyl radicals and HBr, must also approach the diffusion-controlled limit, at low conversion the concentration of molecular bromine will far exceed that of HBr; therefore reaction with bromine will predominate).

Summary data on relative reactivities of alkane C–H bonds towards Cl· and Br· in the gas phase at 298 K are collected in Table 3.4.

Figures 3.3 and 3.4 show that a consequence of skewing our graphs is that, for the exothermic reactions with chlorine, the transition states are reached at points along the reaction co-ordinate not far removed from the starting

Table 3.4 Relative reactivities per hydrogen atom of saturated hydrocarbons towards Cl· and Br·

	Cl·	Br·
CH$_3$–H	0.003	-
RCH$_2$–H	1.0	1.0
R$_2$CH–H	4	82
R$_3$C–H	5.5	*C.*1600

$(CH_3)_3CH + Cl\cdot$. For bromination, they are much further along, and close to the products. The terms 'early' and 'late' transition states respectively are sometimes used. This illustrates what is commonly known as the Hammond Postulate, one statement of which is as follows: 'Both in energy and geometry, the transition states of exothermic reactions are 'reactant-like', whilst those of endothermic reactions are 'product-like'.

The analysis of halogenation presented here places the intuitive (inverse) relationship between reactivity and selectivity mentioned in Chapter 1 on a more secure footing. Indeed, a very simple relationship between activation energy, E_a, and heat of reaction, ΔH, is implied by the close similarities between the plots in Figs. 3.2 – 3.4. This was expressed many years ago in the Evans–Polanyi Equation (3.3), in which E_o and α are constants.

$$E_a = E_o + \alpha\Delta H \qquad (3.3)$$

Unfortunately, not all hydrogen-atom transfers fit a single relationship of this kind. Instead, it is found that different values of the two constants are appropriate to different sets of closely related reactions. Moreover, it might with some justification be argued that hydrogen abstraction by halogen should not be taken as prototypical for the purposes of our discussion. This is because in these reactions polar effects stabilise the transition states and thus facilitate reaction. Recalling the relative stability of a tertiary alkyl *cation*, and the electronegativity of the halogens, this stabilising effect is nicely illustrated by abstraction of tertiary hydrogen from 2-methylpropane by a halogen atom (Fig. 3.5). The effect is also present, albeit less markedly so, for abstraction by halogen from other sites in aliphatic hydrocarbons.

It will be recalled that the slightly *endothermic* reaction of a chlorine atom with methane has an activation barrier of only 15 kJ mol^{-1}. For comparison, the best estimate for the activation barrier to the *thermoneutral* reaction of methyl radical with methane ($CH_3\cdot + H{-}CH_3 \rightarrow CH_3{-}H + CH_3\cdot$), where there is no polarisation of the transition state, is almost 60 kJ mol^{-1}. The two situations are compared graphically in Fig. 3.6. It is, in fact, very fortunate that for reactions with non-polar transition states there is an appreciable activation barrier. Were this not so, one could expect hydrogen exchange between alkyl radicals and alkanes to occur at, or close to, the diffusion-limited rate, so that facile bimolecular radical 'scrambling' would occur during many radical reactions of aliphatic species!

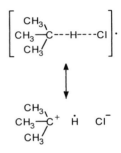

Fig. 3.5 Representation of the transition state for abstraction of tertiary hydrogen from 2-methylpropane by Cl$\cdot$.

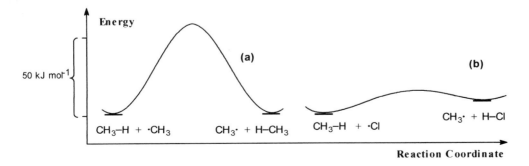

Fig. 3.5 Comparison of reaction profiles for abstraction of hydrogen from methane by CH$_3\cdot$ **(a)** and Cl$\cdot$ **(b)**.

Table 3.5 Dissociation energies of C–H bonds in some substituted methanes

Methane derivative (Y-CH$_3$)	C–H b.d.e. (kJ mol^{-1})
H-CH$_2$–H	439
CH$_3$O-CH$_2$–H	389
N≡C-CH$_2$–H	389
CH$_3$CO-CH$_2$–H	385

A special circumstance arises when both electron-withdrawing *and* electron-donating substituents are attached to the same radical centre. There is evidence that radical stabilisation in such cases is greater than might have been expected. This is a consequence of a special synergistic effect and is referred to as 'captodative stabilisation'. It may be important in peptide-derived radicals (below) which play a role in certain enzyme mechanisms.

RHN–CH·C(=O) NHR'

The ease of hydrogen abstraction from the methyl group in a series of *substituted* toluenes by a radical centred on an electronegative atom – for example *t*-butoxyl – reveals a very good linear free energy correlation with the σ$^+$ substituent constant. Once again, this reflects the importance of polarisation effects in the transition state; e.g.

Table 3.5 gives C–H bond dissociation energies for selected methane derivatives, YCH$_3$, where bond-weakening relative to methane itself provides an indication of radical stabilisation by the substituent. However, it will be realised that any discussion of hydrogen abstraction from these species by some radical X· must take into account not only the relative strengths of the X–H and YCH$_2$–H bonds, but also of the influence of both X and Y on the dipolar character of the transition state. The importance of a cyano group in stabilising an adjacent alkyl radical centre is evident from the data on azo-compound decomposition (p. 16); this is reinforced by the data in the table. However cyano is also strongly electron-withdrawing, so that acetonitrile, CH$_3$CN, is rather unreactive towards hydrogen abstraction by electronegative chlorine atoms. Quite clearly our intuitive ideas of relating selectivity to reactivity based on heats of reaction apply only in strictly comparable circumstances.

Relatively non-polar Y-groups which provide conjugative stabilisation to the radical YCH$_2$·, such as vinyl or phenyl, result in appreciably enhanced rates of hydrogen abstraction, irrespective of the character of X·.

Among recent attempts to incorporate polar and other effects into a modified Evans–Polanyi Equation, the relationship of Eqn 3.4 has been found to give an excellent correlation between the activation barriers for at least 60 different hydrogen-transfer reactions involving a wide range of abstracting radicals and substrates – generalised as A· + H–B → A–H + B· .

$$E_a = E_o f + \alpha\Delta H(1 - d) + \beta(\Delta\chi)^2 + \gamma(s_A + s_B) \qquad (3.4)$$

In this equation, β and γ are new constants, whilst the remaining terms have a significance which may be related to physical characteristics of the reaction partners. Thus $\Delta\chi$ is the difference between the Mulliken electronegativities of A and B, which directly allows for polar effects. Values of *f* depend upon the absolute values of the strengths of the bonds being made and broken. This makes provision for the intuitive belief that two thermoneutral reactions would not have the same activation energy if the strengths of the bonds being transformed were very different. Values of *d* allow for changes in conjugative stabilisation on moving from fully bonded reactant to transition state, as for example in removing benzylic hydrogen from toluene. In a somewhat similar manner, the values of *s* are adjusted to take account of changes in the geometries of A· and B· during progress to or from the transition state. For example, a planar methyl radical becomes pyramidalised as it associates with a fourth hydrogen atom.

This brief discussion of Eqn 3.4 provides only a glimpse of a rather complex empirical relationship, the elegance of which lies in the easily conceptualised significance of the various parameters.

One final but very important aspect of these reactions, implicit e.g. in structure **3.10** (see Fig. 3.2), is that theory suggests that the transition state for hydrogen-atom transfer is co-linear in its preferred geometry. This has obvious implications for intramolecular hydrogen-atom transfer. Thus, even for the six-membered ring transition states of 1,5-hydrogen-atom shifts, which we have seen to be favoured, there is likely to be an element of angle-distortion (Fig. 3.6). 1,4-Shifts are relatively uncommon, and intramolecular 1,3- and 1,2-shifts of hydrogen in radical reactions are virtually unknown.

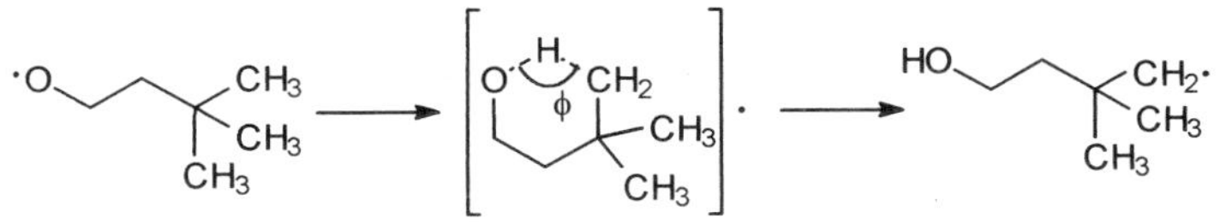

Fig. 3.6 Intramolecular hydrogen transfer in the 3,3-dimethylbutoxyl radical. At the transition state, the angle ϕ must be bent from 180° to minimise angle strain elsewhere in the structure

Although the above discussion has been confined to hydrogen-atom transfer, similar considerations will apply to many other atom-transfer reactions. Of particular importance in preparative chemistry are the transfers of halogen (except fluorine) from tetrahalomethanes (e.g. p. 26, Table 2.2), and from carbon to tin (e.g. p. 9). Also noteworthy, in the context of certain specialised applications (p. 64), are the very high rates of transfer of iodine between simple alkyl radicals.

Polarity reversal catalysis. Unfortunately, the versatility of trialkyltin hydrides as reducing agents is offset, in terms of possible commercial applications, by the toxicity of tin compounds, the last traces of which may be very difficult to remove. Therefore alternative means of bringing about similar reductions have been sought. Moving around the periodic table might suggest silicon hydrides, but the Si–H bond is stronger than Sn–H, and reaction with alkyl radicals is generally too slow for this to be a practical proposition. On the other hand halogen transfer from carbon to a silyl radical *is* very fast. One silicon derivative that is an effective replacement for the tin compounds, however, is tris(trimethylsilyl)silane. The trimethylsilyl substituents are electron-donating compared with alkyl, rendering the central silicon unusually electropositive, and thereby weakening the Si–H bond. The result is an effective and non-toxic alternative to tributyltin hydride.

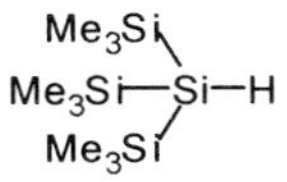

Whilst tris(trimethylsilyl)silane, an example of the application of the use of which will be found in the following section, has proved to be a valuable reducing agent, another ingenious solution to some of the problems associated with the tin hydrides has been proposed which depends on a phenomenon designated 'polarity reversal catalysis'. Where a simple trialkylsilane generally fails as a reducing agent for alkyl halides or chalc-

Tris(trimethylsilyl)silane: the Si–H bond strength is *c.* 330 kJ mol^{-1}, which should be compared with the strengths of other Si–H and Sn–H bonds given in Table 3.3.

ogenides, the presence of a catalytic amount of *t*-butyl mercaptan (2-methyl-2-propanethiol; Me_3CSH) dramatically alters the situation.   The chain-propagating steps involved are depicted below.   Although the S–H bond is only a little weaker than Si–H, hydrogen abstraction from a thiol by a nucleophilic alkyl radical is facilitated by the electronegative character of sulphur, so that the reaction is extremely rapid.   In a similar fashion, the electrophilic thiyl radical rapidly removes hydrogen from the electropositive silicon, regenerating the thiol catalyst.   The sequence is completed by rapid transfer of halogen from carbon to silicon.   The overall result is an efficient replacement of X by H using an inexpensive trialkylsilane.

$$R\cdot \; + \; H\!-\!SBu\text{-}t \longrightarrow R\!-\!H \; + \; \cdot SBu\text{-}t$$

$$R_3Si\!-\!H \; + \; \cdot SBu\text{-}t \longrightarrow R_3Si\cdot \; + \; H\!-\!SBu\text{-}t$$

$$R_3Si\cdot \; + \; X\!-\!R \longrightarrow R_3Si\!-\!X \; + \; R\cdot$$

(Typically, X = Br, I, or SePh)

An equally dramatic demonstration of this effect is found when the *electrophilic t*-butoxyl radical appears to react selectively at an electron-poor C–H bond promoted by a catalyst such as $Me_3N^+\!-\!B^-H_3$.   In this case, the abstracting radical is in fact the strongly nucleophilic $Me_3N^+\!-\!B^-H_2\cdot$.   An ingenious extension of this chemistry has been the accomplishment of a measure of enantioselectivity using analogues of the aminoborane catalyst in which one of the borane hydrogens has been replaced by a chiral substituent. Although extensive treatment of recent adventures into the development of stereoselective radical reactions is beyond the scope of the present text, some further discussion will be found in the final section of this chapter (p. 51).

Radical addition reactions.   When a radical X· adds to an alkene, a σ-bond is made and a π-bond is broken, but in this case it is not a straightforward matter to relate the strengths of these bonds to the heat of reaction in the way that we have done for atom transfer.   First of all, there is the question of how, precisely, does one associate a bond dissociation energy with a π-bond, and secondly, how does one measure the strength of the X–C σ-bond in the new alkyl radical?   However, the actual heats of reaction *are* amenable to study, as are the activation parameters for addition.   Data for the addition of methyl to ethene (ethylene) are illustrated in Fig. 3.7.   As expected, based on a comparison with C–H bond strengths, other alkyl radicals normally react less readily, with *t*-alkyl being the least reactive.

Additions to substituted alkenes, e.g. 2-methylpropene (Eqn 3.5), usually occur at the less substituted carbon.   At first sight, this might appear to be associated with the fact that a more stable radical results, but in fact the rates

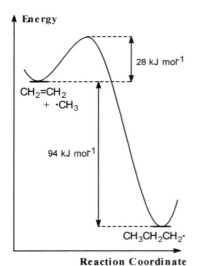

Fig. 3.7 Reaction profile for addition of a methyl radical to ethene

$$R\cdot \; + \; CH_2\!\!=\!\!\underset{CH_3}{\overset{CH_3}{C}} \longrightarrow \underset{CH_3}{\overset{R}{CH_2\!-\!C\cdot}} \tag{3.5}$$

are little different from those of the corresponding reactions with ethene. For example, the measured activation barrier for addition of methyl to 2-methyl-propene (29.5 kJ mol^{-1}) is only marginally different from that for addition to ethene (28 kJ mol^{-1}; see Fig. 3.7), so that steric inhibition of reaction at the more substituted carbon appears to be the determining factor. The detailed picture is somewhat more complicated. For example, alkene substituents will affect the stability of the product radical, but they will also affect the strength of the double bond.

For the addition of methyl to a series of substituted ethenes, XCH=CH$_2$, in which the substituents included CN, NH$_2$, F, and CHO, an excellent correlation has been calculated using a very high level theoretical procedure, between the heat of reaction and the activation barrier. These calculations failed to reveal any dependence on the electron-withdrawing or -donating character of X. Nevertheless, experimental data (see p. 62) support the view that methyl radicals are in fact weakly nucleophilic in their additions to alkenes. Other primary alkyl radicals are considered to exhibit appreciable nucleophilic character and this effect becomes more pronounced with secondary and tertiary radicals.

This nucleophilic character has some interesting synthetic applications. For example, whilst alkyl radical addition to simple alkenes is generally complicated by telomer formation and by other competing side reactions, good yields of 1:1 adducts can be obtained from additions to electron-deficient alkenes such as acrylonitrile. Thus, with tributyltin hydride as a reducing agent, bromocyclohexane and acrylonitrile give **3.11** in good yield (80%; with the iodo-compound, this rises to 95%). A further polar effect operates here: the cyanoalkyl radical is resonance stabilised, but, by virtue of

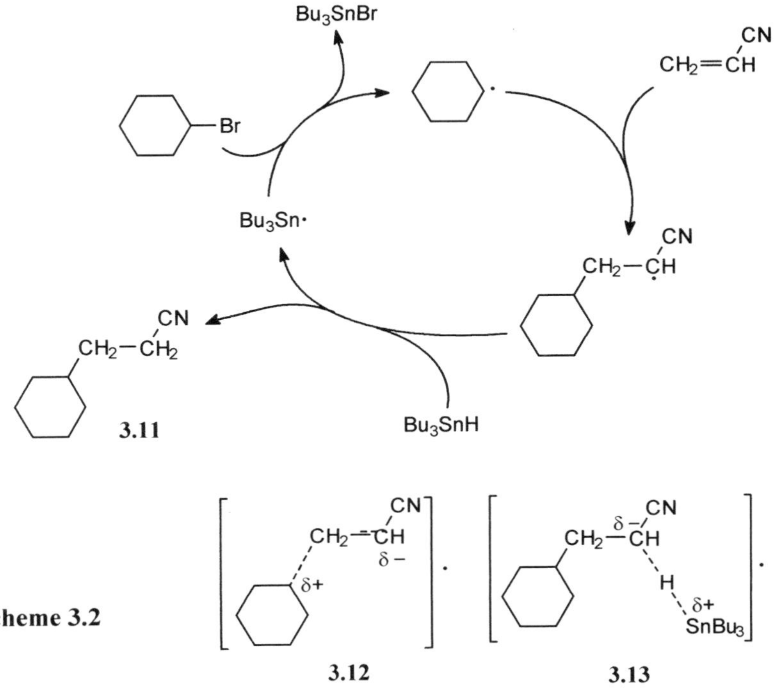

The distinction between methyl and other primary alkyl radicals in respect of nucleophilic character can presumably be associated with the very much higher ionisation potential of the former. For methyl and ethyl the values are *c*. 970 and 810 kJ mol^{-1} respectively.

Scheme 3.2

the cyano substituent, it is now electrophilic and hydrogen abstraction from (electropositive) tin is thereby facilitated. The chain-propagating sequence is displayed in Scheme 3.2, in which an alternative, though self-explanatory, format to the usual sequence of linear reactions has been employed. Also shown, as structures **3.12** and **3.13**, are representations of the two critical transition states, in which the important dipolar character is emphasised. As already noted, in both cases this depends on the electron-withdrawing character of the cyano group.

A synthesis of the macrocyclic lactone zearalenone (as its bis-methyl ether, **3.14**) similarly appears to depend on addition of a nucleophilic carbon-centred radical to an electron-poor alkene, as shown. This is especially remarkable because the intermediate radical (**3.15**) will have reduced reactivity as a consequence of pronounced allylic stabilisation. More pertinent to the present discussion, however, is that, in the absence of the ketone carbonyl, no cyclic product is obtained.

Note here the use of tristrimethyl-silane as reagent in preference to tributyltin hydride.

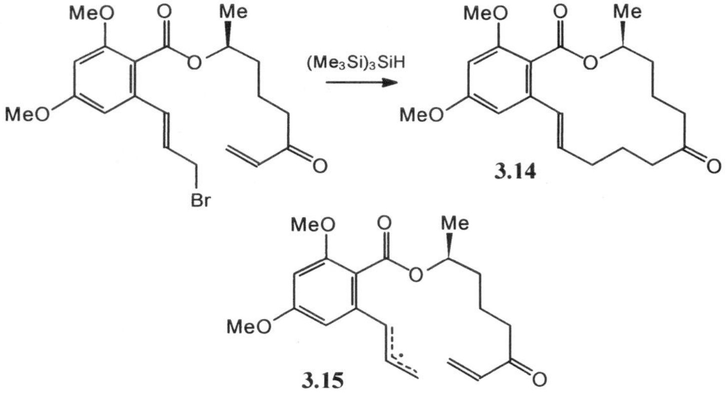

3.14

3.15

It will come as no surprise that polar effects in addition reactions show up during polymerisation, a much-cited example being the co-polymerisation of acrylonitrile and vinyl acetate. The resulting polymer consists of a chain of alternating monomer units. The cyanoalkyl radical, being electrophilic, adds to the electron-rich vinyl acetate. The resulting acetoxyalkyl radical, which is now appreciably nucleophilic, adds to acrylonitrile (Scheme 3.3).

A final example, which incorporates these ideas, but which at the same time nicely encapsulates various other principles from earlier pages, involves the reaction of the peroxide **3.16** with acrylonitrile and α-methylstyrene in methanol solvent, catalysed by an iron salt. A single product, **3.17**, incorporating one molecule of each of the four reactants, was obtained in *c.* 60% yield. The formation of **3.17** is depicted in Scheme 3.4. One-electron reduction of the peroxide gives an alkoxyl radical which ring opens (an intramolecular fragmentation), and the resulting (nucleophilic) alkyl radical adds to acrylonitrile. This forms a radical that now adds preferentially to α-methylstyrene, since this is particularly susceptible to electrophilic attack. The result is a rather stable (tertiary and benzylic) radical that will react only very slowly with alkenes. However, for the first time in the sequence we have a radical which is very susceptible to one-electron oxidation –

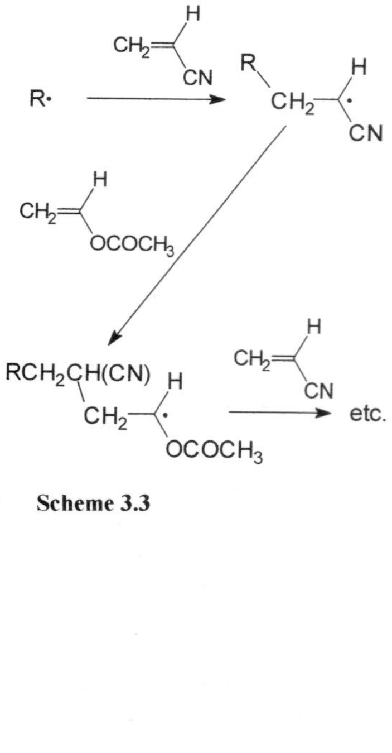

Scheme 3.3

Scheme 3.4

yielding a tertiary benzylic cation – which is immediately quenched by a molecule of methanol. The obvious oxidising agent is iron(III), as shown; this regenerates the initial iron(II) required for peroxide decomposition. Alternatively, it is possible that the chain may be carried by oxidation of the benzylic radical by the peroxide itself.

Some familiarity with mechanism in non-radical chemistry may have acquainted the reader with the direction of approach of a nucleophile to the carbon atom of a carbonyl group. This is perpendicular to the plane defined by the carbonyl and its substituents, and involves attack some 17° (the so-called 'Dunitz angle') away from the vertical and remote from the carbonyl oxygen atom (Fig. 3.8a). In frontier orbital terms, this may be viewed as depending on the interaction of a pair of electrons occupying a bonding or, more commonly, a non-bonding orbital on the nucleophile with a lobe of the π^*-antibonding orbital on carbonyl carbon (Fig. 3.8b).

Athough the situation for an odd-electron system may not be quite so clear cut, radical addition to an alkene is generally perceived in a very similar fashion, with emphasis placed on the interaction of the unpaired electron of the attacking radical with one lobe of the π^*-antibonding orbital of the alkene, illustrated for the case of addition of a methyl radical in Fig. 3.9. This geometry is fully supported by high-level *ab initio* molecular-orbital calculations. A very important consequence is the preference for 5-hexenyl

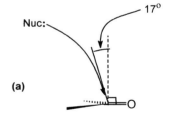

(a)

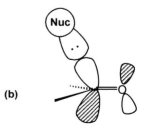

(b)

Fig. 3.8 Approach of a nucleophile towards carbonyl carbon (see text).

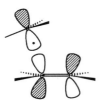

Fig. 3.9 Approach of a methyl radical to an alkene.

3.18

3.19

radicals to cyclise into cyclopentylmethyl rather than cyclohexyl (see p. 13), despite the fact that the latter radical is the more stable. (Remember the relative ease of formation of primary and secondary radicals, and the lower conformational strain in cyclohexane when compared with cyclopentane). Critically, for the transition state leading to the cyclohexyl radical, there is a significant development of ring strain if the proper overlap geometry is to be approximated. In contrast, relatively little build-up of strain is required to reach the transition state for five-membered ring formation. Interestingly, this applies to two quite distinct transition-state conformations, corresponding to 'chair' and 'boat' geometries (**3.18** and **3.19** respectively). There is experimental evidence for involvement of both of these in cyclisations of variously substituted hexenyl radicals.

The distinction between the 5-*exo* and 6-*endo* modes of cyclisation of the 5-hexenyl radical (leading respectively to cyclopentylmethyl and cyclohexyl) is most readily perceived by examination of models.

The influence of proper orbital overlap geometry on the development of the transition state, as illustrated here, is embraced by the term 'stereoelectronic effect' (although carbohydrate chemists, in particular, have used this for a more specific interaction involving lone-pair electrons on oxygen).

In Fig. 3.7, addition of $CH_3\cdot$ to ethene was seen to be appreciably exothermic, and thus far our discussion of radical addition to alkenes may seem to have implied that such additions are always irreversible. That this is clearly not the case, however, was exemplified for bromine-atom addition in our description of the mechanism of allylic bromination by NBS (p. 25). There are now many instances in which the reverse of addition (i.e. fragmentation) has found application in synthetic strategies. In the example of Scheme 3.5, which achieves *C*-allylation in an efficient radical chain reaction, tributyltin radical effects S_H2 displacement from selenium; the resulting alkyl radical adds to allyltributylstannane, whereupon the adduct **3.20** fragments, regenerating the tributyltin radical. The sequence is initiated by ultraviolet irradiation.

Scheme 3.5

Initiator thermolysis. Polar effects may also be identified in the rates of thermolysis of peroxides. Attention was drawn, in Chapter 2, to the evidence that the decomposition of certain peroxides, such as peroxyester **2.14**, involves simultaneous rupture of more than one bond. When the rate of this decomposition is compared with the rates of decomposition of the corresponding *p*-methoxy- and *p*-nitro-compounds, it is found that there is a marked substituent dependence which may most readily be rationalised in terms of a dipolar contribution to the transition state (*cf.* **3.21**). The relative rates of unimolecular decomposition of the three peroxyesters are compared in Table 3.6.

3.21

Table 3.6 Relative rates of decomposition of *para* substituted *t*-butyl phenylperoxyacetates

Peroxyester	Relative rate (80°C)
p-NO$_2$-C$_6$H$_4$CH$_2$CO$_2$OBu-*t*	0.15
C$_6$H$_5$CH$_2$CO$_2$OBu-*t*	1
p-MeO-C$_6$H$_4$CH$_2$CO$_2$OBu-*t*	100

Initiator thermolysis has also been used to probe aspects of radical structure. For example, azoisobutane (**3.22**; see also p. 16) decomposes significantly more slowly than the azoadamantane **3.23** or the azobicyclo-octane **3.24**. This presumably reflects the preference for the incipient tertiary alkyl radicals to favour an essentially planar structure – which is not possible for either of the two bridgehead systems. (See alsothe side note on p. 32.)

3.22   **3.23**   **3.24**

In another example, relative rates of decomposition of analogues (**3.25**) of the azo-compound AIBN (**2.16**), should reveal information about the ease of formation, and hence stability, of the derived radicals ·C(CN)R$_2$. When the relevant rate comparisons were carried out with R$_2$CCN = α-cyanocycloalkyl (**3.26**), the results (Table 3.7, overleaf) were interpreted in terms of the constraints on the change of hybridisation at carbon from *sp*^{3} to *sp*^{2} as the cyanoalkyl radicals are formed. This is sometimes referred to as internal strain or 'I strain'. Compared with the reference point of n = 6 (i.e. α-cyano-cyclohexyl), it is not surprising to find an appreciable rate retardation in the cyclobutane case (n = 4), since more strain must be present in a small ring containing an *sp*^{2} carbon than in one where the corresponding carbon is tetrahedral. It is perhaps more interesting to note the significant rate *acceleration* that appears when n = 5, or especially 7, 8, or 10.

3.25

3.26

In all of these cases of azo-compound thermolysis, symmetry of the transition state for decomposition is assumed. This would preclude, or at least minimise, the complicating influence of polar effects.

Table 3.7 Relative rates of decomposition of azo-compounds **3.26** as a function of ring size

n in **3.26**	Relative rate of decomposition
4	0.025
5	12
6	1
7	190
8	1400
10	320

Intramolecular reactions. One feature exemplified by cyclisation of hexenyl radicals is of far-reaching importance. It relates to a comparison of the rate of intramolecular addition to that of a suitable intermolecular model system. Since radical addition to 1-alkenes occurs at the terminal carbon, a suitable model for comparison with hexenyl itself is not readily accessible. However, rate data for the two systems shown in Scheme 3.6 *are* available.

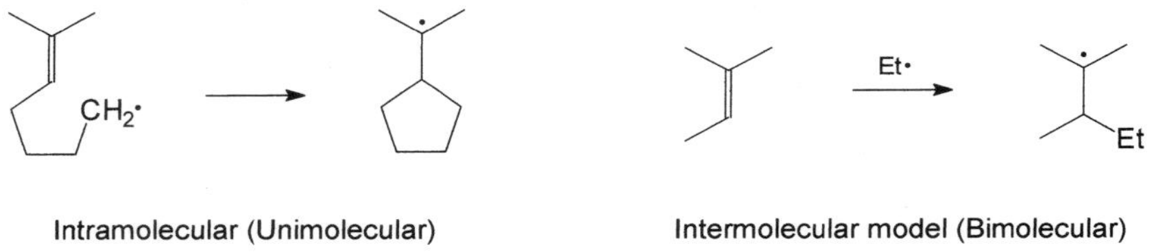

Intramolecular (Unimolecular)   Intermolecular model (Bimolecular)

Scheme 3.6

We first have to examine how to make a sensible numerical comparison between the rate of a unimolecular cyclisation and that of a suitable bi-molecular model. Consider the circumstance that the bimolecular reaction is carried out under pseudo-first order conditions, in which a low concentration of radical interacts with molar (1 M) alkene. Suppose that it is found that the pseudo-first order rate of this model is exactly equal to the true first-order rate of the cyclisation. We might then reasonably say that the 'effective concentration' (or 'effective molarity') of alkene in the intramolecular cyc-lisation reaction is 1 M. Should the intramolecular reaction be ten times more rapid than the intermolecular model, then the effective concentration of the alkene in the intramolecular process is 10 M, and so on.

Although the concept of effective molarity (E.M.) is most easily visualised in this way, it is properly defined as the ratio k_1 / k_2, where k_1 and k_2 are the true first-order and second-order rate constants for the intramolecular and bimolecular-model reactions respectively.

One might anticipate that a maximum E.M. would be of the order of 10 – 20 M, corresponding to the solvent being the substrate. But when the rate data for the reactions in Scheme 3.6 are compared, we find that the effective concentration of alkene for the intramolecular case is *c.* 10^5 M. That such large numbers are possible is widely considered to reflect the entropic advantage of the fewer degrees of freedom for the intramolecular process. It is analogous to the chelate effect in inorganic chemistry. Values of E.M. approaching 10^8 have been encountered in, for example, intramolecular nucleophilic reactions of carbonyl compounds. But perhaps the most important manifestation of the effect is in intracomplex processes of enzyme-catalysed reactions. Returning, however, to radical chemistry, the effect of intramolecularity appears to allow certain synthetically useful reactions to proceed in situations where no intermolecular counterpart has been observed. Three examples, each of which involves 5-*exo* cyclisations akin to that of the 5-hexenyl radical, are given in Scheme 3.7. The first case, using a tin hydride reduction of a hydroxybutyl halide, affords a reasonable yield of cyclopentanol. Here, despite the significant kinetic advantage arising from intramolecular addition to carbonyl carbon, the equilibrium constant for the cyclisation is only *c.* 0.002. But the equilibration is very rapid, and the reaction of tributyltin hydride with the alkoxyl radical is 1000 times faster than it is with the uncyclised radical (**3.27**). This is sufficient to offset the unfavourable position of the cyclisation equilibrium. For the second example, the cyclisation is preserved by rapid (possibly concerted) loss of molecular nitrogen.

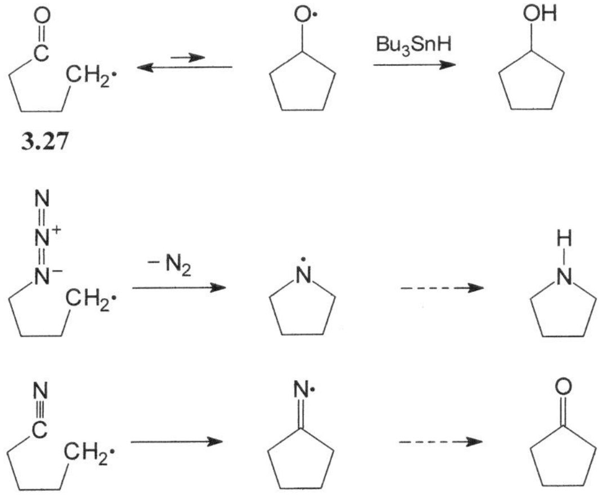

Scheme 3.7

Finally, it is important to note that similarly dramatic intramolecular effects are *not* generally encountered in atom-transfer reactions, for which E.M. values of 1-10 M are more typical. One possible explanation is the difficulty of attaining the ideal co-linear geometry for atom transfer (see Fig. 3.6 and the associated discussion).

The persistent radical effect. One more kinetic aspect of radical chemistry is deserving of consideration. Under certain circumstances, predicted radical-coupling products are not found, and formation of others is unexpectedly efficient. Literature examples, which have appeared at regular intervals since the 1950's, have frequently been accompanied by suggestions of some inexplicable 'preference for unsymmetrical radical combination', or, more

3.29

Dimerisation and disproportionation
reactions of the cyclohexyl radical.

dramatically, protestations that free radicals cannot possibly be involved. Illustrative of these apparent anomalies are the products formed during the thermal decomposition of the unsymmetrical azo-compound Ph–N=N–CPh$_3$ in a hydrogen-atom donor solvent such as cyclohexane. The azo-compound decomposes into nitrogen, Ph·, and ·CPh$_3$. The organic products (> 95%) are benzene and (triphenylmethyl)cyclohexane (**3.28**). No detectable quantities of cyclohexene or bicyclohexyl (**3.29**) are formed, although these are known products of other reactions involving cyclohexyl radicals. The decomposition is cleanly first order; there is no evidence of a chain reaction.

The explanation depends on the persistence of ·CPh$_3$, and on the fact that the reaction of cyclohexyl with ·CPh$_3$ is as fast as its dimerisation or disproportionation. This means that the fate of the cyclohexyl radicals depends on the relative concentrations of the two radical species. Consider the first moments of reaction. Phenyl and triphenylmethyl are formed in equal proportions. The phenyl is exceptionally reactive, and is rapidly replaced by cyclohexyl. This is removed by termination reactions, and, to the extent that these involve two cyclohexyl radicals (*i.e.* dimerisation and disproportionation), there will be an equivalent build up of triphenylmethyl radical in equilibrium with its dimer. Very rapidly, therefore, the concentration of triphenylmethyl will greatly exceed that of cyclohexyl – by several orders of magnitude – but yet will remain at microscopic levels. Any newly formed cyclohexyl will now be scavenged by ·CPh$_3$. But this does not deplete the 'reservoir' of ·CPh$_3$, since every cyclohexyl radical to be formed is accompanied by another new triphenylmethyl.

The rather dramatic consequences of the persistent radical effect have obvious importance in synthesis. Thus NO is the persistent radical which accounts both for the efficiency of the Toray process (see p. 23), and for that of many synthetic reactions involving photolysis of organic nitrites, exemplified opposite. Likewise, Fremy's radical plays a similar role in the Teuber Reaction (p. 8).

Fremy's radical, ·CPh$_3$ and NO may all be thought of as 'stable' radicals, but the effect is also operative for other persistent species whose bimolecular self-reaction is significantly slower than the diffusion-controlled limit but which will eventually decay irreversibly. Alkylperoxyl radicals (ROO·) fall into this category (see Chapter 5), in which context confusion has arisen recently in the study of model systems designed to mimic the enzymatic (*non*-radical) oxidation of aliphatic hydrocarbons. Experimental results were initially interpreted as having succeeded in reproducing the enzyme mechanism, precisely because certain characteristically free-radical products were absent. However, the correct interpretation is now believed to depend on involvement of alkylperoxyl radicals behaving as persistent radical intermediates.

Medium effects. There are some interesting exceptions to the generalisation (p. 29) that reactions involving electrically neutral radicals are insensitive to solvent variation. One example of this is encountered in a system which was the subject of fairly detailed analysis earlier in this chapter, namely alkane chlorination. The relative reactivities of primary, secondary and tertiary hydrogens towards chlorine atoms found in the gas-phase (see Table 3.4, p. 36) are very little altered when measured in solution in CCl_4. In contrast, when the solvent is benzene, the tertiary:primary ratio rises (at 298 K) to 32. In carbon disulphide it is 225. These variations are attributable to complexation of the chlorine atom in the latter two solvents, and a concomitant reduction in its reactivity and increase in selectivity. The complexation of chlorine atoms with benzene has been directly observed spectroscopically in flash photolysis studies.

A somewhat surprising solvent effect has been detected in the fragmentation of *t*-butoxyl radicals. In early investigations of relative hydrogen donor reactivity of a series of substrates, R–H, towards butoxyl, comparisons had been made on the basis of the relative yields of acetone and *t*-butanol. Under comparable conditions, the more butanol, the higher the reactivity of R–H (Scheme 3.8). Used quantitatively, this makes the assumption that the rate of the fragmentation reaction is essentially independent of solvent. In the majority of circumstances this is probably justified, but even in the early work there was evidence that the fragmentation reaction was appreciably

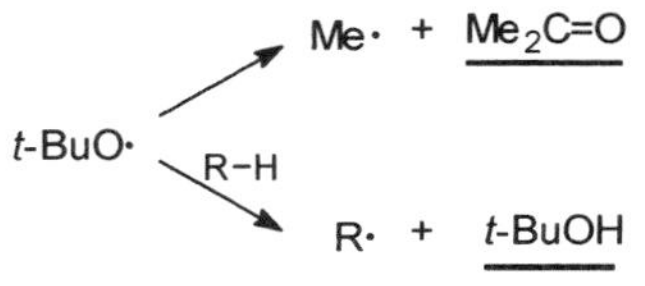

Scheme 3.8

accelerated in polar solvents. This has been put on a more quantitative footing with a careful study of the fragmentation of cumyloxyl radicals ($PhCMe_2O\cdot$) into methyl and acetophenone. This is approximately seven times faster in acetic acid than in CCl_4, whilst the rate of abstraction of hydrogen from cyclohexane in these solvents is essentially the same.

In a range of solvents, the variation of fragmentation rate of cumyloxyl shows a good correlation with solvent-polarity parameters. This shows that the acetic acid result is not a consequence of hydrogen bonding. In contrast, hydrogen bonding clearly does influence reaction rates in certain other circumstances, for example hydrogen abstraction from the OH group of phenols. From Table 3.3 (p. 34) the O–H bond strengths of phenols are seen

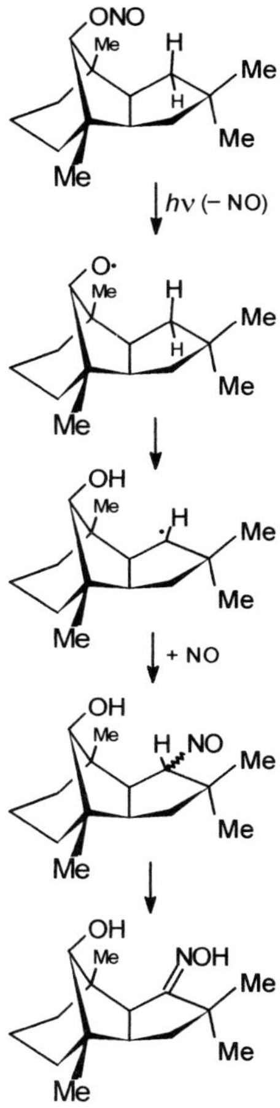

In the above example, based on the ring-system of the sesquiterpene caryophyllene, the (racemic) product was obtained in excellent yield. Note the rearrangement of the secondary *C*-nitroso-compound into an oxime. Essentially the same chemistry was deployed in the synthesis of aldosterone (see p. 13).

Another circumstance in which hydroxylic solvents, especially water, can promote a marked deviation from expected behaviour, is found in the occurrence of an apparent 1,2-shift of hydroxyl, noted in observations of the alkoxyalkyl radical **3.30**. Closer study revealed that this is an acid-catalysed process, proceeding *via* the radical cation **3.31**. More recently it has become apparent that rather similar processes are implicated in the mechanisms of some very important enzyme-catalysed transformations.

$$\overset{\bullet}{MeOCHCH_2OH} \longrightarrow MeO\overset{\overset{\displaystyle OH}{|}}{C}HCH_2\bullet$$

3.30

$$MeO\overset{+}{=}CH{-}CH_2\bullet$$

3.31

with H$^+$, $-H_2O$ going down-left; H_2O, $-H^+$ going up-right.

A long established synthetic sequence, which depends on intramolecular *hydrogen abstraction* by a protonated aminyl, is the Hoffman-Löffler-Freytag reaction. Photolysis of a protonated *N*-chloramine generates the amine radical cation which abstracts hydrogen from the δ-carbon. A chain reaction ensues, in which the alkyl radical accepts chlorine from the protonated chloramine. Basification yields a pyrrolidine derivative. An example showing propagating steps and product formation appears below.

to be lower than those for alcohols, reflecting electron delocalisation in the derived phenoxyls. This weakness shows up in a wide variety of phenol oxidation reactions (e.g. the Teuber reaction, p. 8), many of which are of biological importance (see, for example, p. 88). But the rate of hydrogen abstraction from a given phenol may be appreciably reduced on changing from e.g. a hydrocarbon solvent to one capable of acting as a hydrogen-bond acceptor. With cumyloxyl, phenol itself reacts more than 100 times more slowly in acetonitrile than it does in CCl$_4$. Effectively, it appears that hydrogen-bonding to solvent seems to strengthen the O–H bond. More careful kinetic studies have revealed, however, that the proper explanation is that reaction occurs only with that fraction of the phenol which is *not* H-bonded. Since the concentration of this is greatly reduced in acetonitrile, the overall reaction appears to be slowed.

An alternative form of molecular association to hydrogen bonding is that with Lewis acids. Indeed, in the case of protonation, we are moving into the realms of radical-ion chemistry, addressed in Chapter 6. However, a few examples will be presented here, not least because of the dramatic recent developments in the employment of Lewis-acid complexation to achieve stereocontrol, briefly examined in the concluding Section of this Chapter.

Aminyl radicals, RR'N·, are relatively unreactive species, but the nitrogen lone pair retains its basic character, and it has been possible, using flash photolysis and kinetic spectroscopy to measure the rate of intramolecular addition of the aminyl radical **3.32**, and to compare it with the values for the BF$_3$-complexed, and protonated analogues, with the results indicated. Clearly, coordination of the nitrogen lone pair generates a more reactive radical species.

One effect of Lewis-acid complexation in the aminyl system is to render the radical centre more electrophilic. A dramatic instance of the application of this to a carbon-centred radical involves the radical addition reaction shown in Eqn 3.6. It is known that an electron-withdrawing substituent will make an alkyl radical weakly electrophilic (*cf.* for example, Scheme 3.3), but despite this, radical-chain addition of RCOCH$_2$Br to an internal alkene is an inefficient process. The example of Eqn 3.6, with *N*-(α-bromoacetyl)-oxazolidinone, illustrates how complexation to the carbonyl groups can alter matters. Initiated by Et_3B/O_2 at 25°C, and using five equivalents of *cis*-3-hexene, the yield of 1:1 adduct is below 10%. But with one equivalent of $Yb(OSO_2CF_3)_3$ present, the addition is essentially quantitative.

$$(3.6)$$

The use of triethylboron as an initiator is applicable at temperatures down to that of dry ice. A spontaneous autoxidation of the trivalent boron occurs (see p. 53) with traces of adventitious oxygen being sufficient to facilitate this process, and the radicals formed act to initiate the reaction being investigated.

We shall return to Lewis-acid complexation with *N*-acyloxazolidinones in the next Section.

Coordination of a Lewis acid at the site subject to radical attack may also modify radical reactivity. Thus the nucleophilic character of an alkyl radical, which shows up in its tendency to react at an electron-deficient double bond, may be accentuated if the electron-withdrawing group attached to the double bond is co-ordinated to a Lewis acid. An example is found in the improved yield obtained for the difficult reductive alkylation of methyl crotonate (Eqn 3.7) in the presence of 1.5 equivalents of $EtAlCl_2$ (33% compared with <1% in the absence of the Lewis acid).

$$(3.7)$$

(Initiated with Et_3B/O_2 at -78°C)

Steric effects and stereocontrol. One of the most familiar and straight-forward factors in modification of organic reactivity is the steric effect. However, in early observations on radical behaviour this often seemed rather small. For example, in investigations of the dimerisation of radicals $R^1R^2R^3C\cdot$, there was usually a near 50:50 mixture of meso and *d,l* isomers produced. In other respects, too, ideas of stereocontrol seemed fruitless. For example, since an alkyl radical of the form $R^1R^2R^3C\cdot$ is effectively planar at the radical centre, it would be expected to react with equal probability on either face, forming racemic product. Furthermore, when the chiral centre and the radical centre are on separate atoms, appreciable asymmetric induction was seldom observed. This may be attributed, at least in part, to exothermicity of key reaction steps. This, in turn, will normally be associated with early transition states, in the structures of which the reaction partners remain relatively well separated. Therefore minimal steric interaction is unsurprising. Nevertheless, in relatively rigid systems a measure of stereocontrol *can* sometimes be observed. A simple example is the addition of the 2-methoxypentyl radical (**3.33**) to acrylonitrile, in which

More recently, modest diastereo-selectivity *has* been demonstrated in some radical coupling reactions. And the rather dramatic consequences of steric crowding in triisopropylmethyl (p. 33) should not be forgotten. In another interesting study, what might, perhaps, be thought of as an 'external steric effect' has been encountered in the photolysis of a hydrocarbon solution of the *meso* isomer of $PhCMe(CHMe_2)$–N=N–$CMePh(CHMe_2)$ which has been rapidly frozen to a glassy state. Only the *meso* isomer of $PhCMe(CHMe_2)$–$CMePh(CHMe_2)$ is formed, in marked contrast to photolysis in liquid solution, when both stereoisomers are produced.

3.33

ca. 15%

ca. 85%

The *N*-acyloxazolidinones are readily synthesised by acylation of suitable chiral oxazolidinones. The products of the radical chemistry may then be hydrolysed to carboxylic acids using alkaline hydrogen peroxide. Strictly, these radical reactions are diastereoselective. Genuinely enantioselective variants have now been developed using achiral acyloxazolidinones and a Lewis acid which is co-ordinated to a homochiral ligand.

more than 80% of the product is formed *via* addition *trans* to the methoxyl group.

The past decade has seen an explosion of interest in the possibility of exercising more efficient stereocontrol in synthetic radical chemistry, in particular by probing the use of a variety of chiral auxiliaries, which may be removed after the radical reaction has been carried out. The effective result of this strategy is enantioselective chemistry. Some of the most interesting examples employ variants of the acyloxazolidinone chemistry introduced above (Eqn 3.6), where a critically important aspect is chelation of a suitable Lewis acid. Just one instance of the potential of this approach is illustrated below. The reaction is allylation of the enantiomerically pure (i.e. 'homochiral') α-bromopropionamide derivative **3.34**, using allyltributyltin (*cf.* Scheme 3.5, p. 44) at −78°C, initiated by BF_3/O_2. In the uncomplexed reaction, the ratio of diastereomeric products, **A:B**, is 1:1.8. In the presence of magnesium bromide, however, this becomes 100:1. Evidently, chelation of magnesium prevents rotation of the acyl side chain (*cf.* **3.35**), and the bulky diphenyl-methyl substituent then obstructs one face of the radical intermediate, forcing attack on the tin compound from the opposite face.

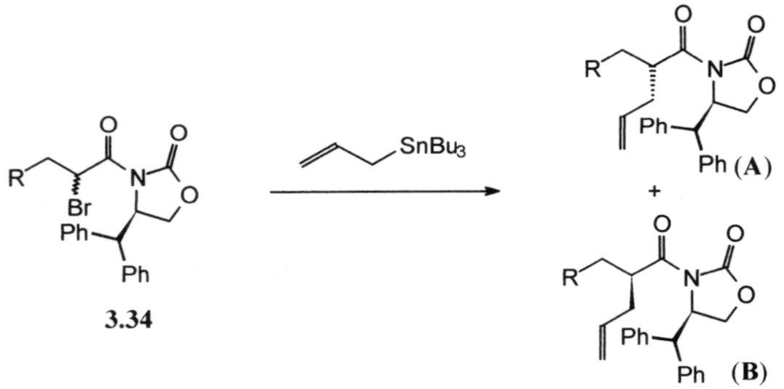

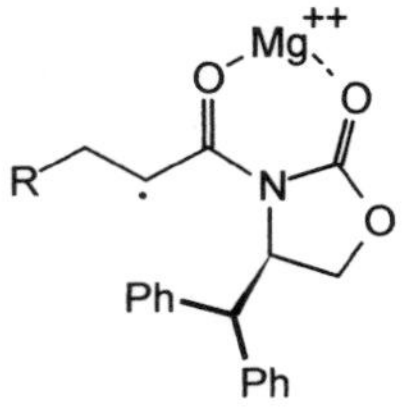

4　Experimental methods

4.1　Detection of radicals

Classical procedures

The earliest methods for indicating the presence of radicals as intermediates
in organic reactions included formation of polymer when an easily polymerised monomer such as styrene was added to the system. However, the results
of such experiments may be quite misleading, since, as we have already seen,
radical reactions of styrene do not always lead to polymer. Instead, as for
example in its reaction with carbon tetrabromide (p. 10), chain transfer,
leading to 1:1 adduct formation, may be preferred.

A superior procedure was to add a stable radical such as DPPH (**1.2**) as a
radical scavenger. The disappearance of the colour of such a species not only
acted as a qualitative indicator of the formation of radical intermediates,
often it would also provide a convenient and quantitative method of
following the rate of homolysis of peroxides or other initiators. Even here,
however, although the scavenger may be capable of trapping another radical
at the diffusion-limited rate, problems can arise. For example, in early
attempts to determine whether the autoxidation of organoboranes has a
radical mechanism, various radical scavengers including the vinyl monomer,
methyl methacrylate; iodine; 1,4-dihydroxybenzene (quinol), and even
DPPH, all failed to inhibit the reaction significantly. However, galvinoxyl
(**1.3**), which is an excellent scavenger for both oxygen-centred and carbon-
centred radicals, did provide dramatic inhibition, and this and subsequent
evidence has established that these oxidations do in fact involve a radical
chain mechanism (Scheme 4.1) which includes, as a key step, an S_H2
displacement of an alkyl radical from boron.

$$R\cdot \ + \ O_2 \ \longrightarrow \ RO_2\cdot$$

$$RO_2\cdot \ + \ R_3B \ \longrightarrow \ R\cdot \ + \ R_2BOOR$$

and:
$$RO_2\cdot \ + \ R_2BOOR \ \longrightarrow \ R\cdot \ + \ RB(OOR)_2$$
$$Etc.$$

Scheme 4.1

In a context somewhat more familiar to the organic chemist, very similar
mechanisms operate in the reactions with molecular oxygen of organolithium
compounds and of Grignard reagents.

Illustrative of other experimental evidence which led to the conclusion that
radicals *are* involved in these reactions are the observations that both *exo-*

and *endo*-2-methyl-2-norbornylboranes **4.1** and **4.2** give identical mixtures of peroxyboron compounds, and that, as the oxygen supply is reduced, the hexenyl Grignard **4.3** gives increasing amounts of product derived from the characteristic ring-closure to cyclopentylmethyl radicals (*cf.* pp. 13, 43-4).

Spectroscopic methods for radical detection

The use of conventional optical (ultraviolet/visible) spectroscopy may be practical for some experiments of the kind outlined above, where moderate concentrations of strongly absorbing species are readily attainable. But it is generally unsuitable for monitoring the much lower concentrations of short-lived reactive species such as simple alkyl radicals (which in any case may have no useful absorption features).

Electron spin resonance spectroscopy. There is, however, a rather sensitive magnetic resonance technique which may be thought of as the unpaired electron counterpart of proton magnetic resonance spectroscopy. Known as electron spin resonance (e.s.r.) spectroscopy, or sometimes (particularly by inorganic chemists) as paramagnetic resonance spectroscopy, it is capable of detecting, and providing structural information on, many paramagnetic species – at concentrations down to 10^{-6} M or lower. Despite the relatively high sensitivity, this is insufficient for direct detection of reactive radicals at concentrations normally encountered in laboratory situations (*cf.* Chapter 3).*
Nevertheless, it is important to present a brief, and essentially phenomenological introduction to e.s.r. methodology, both because of its application to the determination of the kinetics of radical reactions under conditions where higher radical concentrations are being generated, and because of its utilisation in an indirect procedure which has found some use in both chemical and biochemical investigations.

---

*   The interested reader is directed to the bibliography for references to fuller descriptions of instrumentation and spectrum interpretation. Although several magnetic field ranges have been used, the most common is centred on *c.* 0.33 T, at which the resonance condition for an organic radical will be found at a frequency of *c.* 9 GHz (this is in the microwave region of the electromagnetic spectrum, and corresponds to a wavelength of *c.* 3 cm).

Comparison with proton n.m.r. is instructive. In the ^{1}H n.m.r. spectrum of acetaldehyde, the aldehyde proton appears as a 1:3:3:1 quartet. This is a consequence of magnetic interaction with the three methyl protons. The same quartet pattern is observed for the electron in the e.s.r. spectrum of the methyl radical (Fig. 4.1).

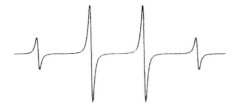

Fig. 4.1 Electron spin resonance spectrum of a methyl radical, showing the quartet splitting arising from magnetic interaction with the three protons.

Usually, e.s.r. spectra are displayed (as in Fig. 4.1) as the first derivative of the absorption. Historically, this is a consequence of the earliest detection procedures, but the format has been maintained in modern instrumentation because many radicals give complicated multi-line spectra, and in these circumstances better resolution is apparent in the derivative format. As an illustration of this, Fig. 4.2 displays the conventional (first derivative) spectrum of triphenylmethyl set beneath an absorption spectrum shown under identical conditions of spectral line shape and line width.

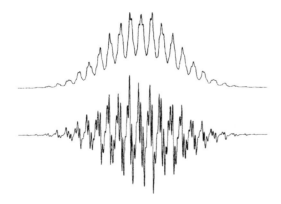

Fig. 4.2 Computer-simulated first-derivative electron spin resonance spectrum of the triphenylmethyl radical set below the absorption mode spectrum computed using identical splitting and line-width parameters.

In proton n.m.r. spectroscopy, as well as simple spin multiplets, multiplets of multiplets may be encountered. For example, the proton on C-2 of *trans*-1,3-dichloropropene would show six lines [a doublet of triplets, resulting from different magnitude couplings to the proton on C(1) and to the two protons on C(3)]. In the e.s.r spectra of radicals, electron delocalisation frequently results in far more complex splitting patterns. The triphenylmethyl radical has a delocalised structure, in which the unpaired electron interacts strongly with three equivalent *para* protons and six equivalent *ortho* protons; it also interacts, but more weakly, with six equivalent *meta* protons. The result is a spectrum comprising a quartet of septets of septets, or $4 \times 7 \times 7$ (= 196) lines. Not all of these are resolved. Many of the smaller ones which occur near to the centre of the spectrum (Fig. 4.2), are hidden under more intense lines.

Spectra obtained from radicals dissolved in mobile solvents are primarily characterised by the splitting pattern and by the magnitude of splittings in individual multiplets. In conventional e.s.r. instrumentation (in contrast to most modern n.m.r. spectrometers), the resonance condition is achieved at a constant frequency, the spectrum being obtained by scanning the magnetic field. The magnitudes of nuclear hyperfine splittings are measured in magnetic field units. Thus for the methyl radical the hydrogen hyperfine splitting (between any two adjacent lines in the quartet) is about 2.3 millitesla (mT). This is given the symbol a_H. Its precise value varies very slightly with both solvent and temperature. Interestingly, a second parameter, the 'g-value', which is the e.s.r. counterpart of chemical shift, tends to vary very little from one organic radical to the next, and is frequently disregarded. Nevertheless, its precise measurement can give structural information, not least since its exact value depends upon spin-orbit coupling, and this can become significant when there is a heavy atom located near to the unpaired electron.

> Variation in *g*-values for organic radicals is usually very much smaller than the nuclear hyperfine splittings, so that in circumstances under which more than one radical is present, their spectra will overlap.

The reader may recall that, in a typical radical reaction in liquid solution, the concentration of radicals may be of the order of 10^{-8} M or even lower, whilst the limit of detection is generally not much below 10^{-6} M. How, then, is it possible to record a spectrum of methyl? Several solutions to this problem have been developed, of which one of the most widely used involves the focusing of light from an intense ultraviolet light source on to a static photolabile sample properly positioned in a temperature-controlled cell in the spectrometer. To obtain the spectrum of methyl, one approach might be to photolyse a solution of diacetyl peroxide in a suitably unreactive solvent. For this purpose, one widely used solvent is liquid cyclopropane, in which the C–H bonds are particularly strong (see side-note on p. 33), although for radicals less reactive than methyl, a range of options is available.

> Vacuum degassed cyclopropane (b.p. -33°C) may be used in a sealed quartz tube at temperatures up to ambient without undue hazard.

Electron spin resonance and radical structure. Not surprisingly, there is a relationship between the distribution of the unpaired electron in a radical and the magnitude of the nuclear hyperfine interactions. The methyl radical is planar, with the unpaired electron in a *p*-orbital on carbon, and, as we have just seen, there is a hyperfine splitting due to hydrogen of *c.* 2.3 mT. Making the assumption that the cyclopentadienyl radical is planar, with the unpaired electron equally distributed between five planar carbon centres, one might naively expect a splitting to each hydrogen of 2.3/5 = 0.46 mT. The actual value is 0.6 mT – modest agreement, from a very simple approach. Indeed, agreement might have been better still had we in some way been able to take account of the fact that the bond angles in the two systems are necessarily different. Simple relationships of the kind implied here have been very useful in estimating the unpaired electron distribution in many planar π-electron systems.

Cyclopentadienyl

It may come as a surprise that in the spectrum of the *ethyl* radical the magnitude of the splitting from the three β-protons of the methyl group is greater than that from the two α-protons attached to the radical centre. In fact, it is easy to see how unpaired electron density associates with the β-protons when the hyperconjugative interaction, depicted in Structure **3.5** (p. 32), is recalled. In contrast, α-protons lie in the nodal plane of the *p*-orbital, so that *any* interaction might at first sight seem surprising. The qualitative picture used to explain the α-splitting recalls the basis for Hund's Rule, which states that in its electronic ground state an atom has that electron configuration with the greatest number of unpaired electrons (with parallel electron spins). By analogy with Hund's Rule, the repulsion between electrons in an alkyl radical is least when the electron spins are parallel. This manifests itself as 'spin polarisation'. In the α-C–H bond pair of electrons, there is a tendency for the electron with the same spin as that of the unpaired electron to reside closer to carbon, and that with the opposite spin to associate with hydrogen (Fig. 4.3). This implies that the interaction with the α-hydrogen is *negative*, whilst the more direct hyperconjugative interaction with the β-hydrogen (see p. 32) is *positive*. The rather curious concept of spin polarisation is fully borne out by a complete theoretical analysis. And although the experimental measurement of the sign of a_H requires special instrumentation, its negative value has been established in suitable instances.

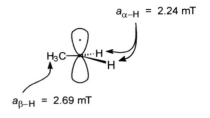

$a_{\alpha-H} = 2.24$ mT

$a_{\beta-H} = 2.69$ mT

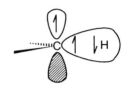

Fig. 4.3 Spin polarisation in the α-C–H bond of an alkyl radical.

One fruitful series of e.s.r. studies has been in the field of S_H2 reactions of polyvalent atoms. In a number of cases, it has been possible to demonstrate that an intermediate is formed. A particularly interesting situation which arises with reactions of trivalent phosphorus compounds is exemplified by the reaction of methoxyl radicals with trimethylphosphine. Electron spin resonance experiments have not only revealed the intermediacy of the hypervalent 'phosphoranyl' radical (**4.4**), but also, using isotopic labelling, have shown that this decays to give methyl radicals both by the simple S_H2 process (path 'a'), and by an alternative fragmentation route (path 'b').

$$MeO\cdot \;+\; Me_3P \;\longrightarrow\; \left[\begin{array}{c} Me \\ | \\ MeO-P-Me \\ | \\ Me \end{array} \right]\cdot \quad \begin{array}{l} \overset{"a"}{\longrightarrow} \quad MeOPMe_2 \;+\; Me\cdot \\[8pt] \underset{"b"}{\longrightarrow} \quad O{=}PMe_3 \;+\; Me\cdot \end{array}$$

4.4

Spin trapping. An indirect procedure by which e.s.r. has been used for monitoring radical participation in chemical and biochemical reactions is called 'spin trapping'. It depends on the use of small quantities of a radical scavenger which is diamagnetic, i.e. it is a non-radical species, but one which reacts especially rapidly with a wide range of radicals to form moderately persistent (paramagnetic) adducts – the so-called 'spin adducts'. This is generalised in Eqn 4.1.

$$\begin{array}{ccc} R\cdot & + & ST \\ \text{reactive} & & \text{spin} \\ \text{radical} & & \text{trap} \end{array} \longrightarrow \begin{array}{c} (R\text{-}ST)\cdot \\ \text{spin} \\ \text{adduct} \end{array} \qquad (4.1)$$

The relatively slow rates of disappearance of the spin adducts allow their concentrations to build up to levels well above the detection limits of the spectrometer. In favourable circumstances the spectra then permit the trapped radicals to be characterised and other, e.g. mechanistic, deductions to be made.

Two classes of compound in particular have found application as spin traps. Both of these form spin adducts belonging to the group of radicals known as nitroxides (*cf.* di-*t*-butyl nitroxide, p.1). These are *C*-nitroso-compounds, and nitrones, represented here by 2-methyl-2-nitrosopropane (MNP; Eqn 4.2) and the cyclic nitrone, 5,5-dimethyl-1-pyrroline-*N*-oxide (DMPO; Eqn 4.3) respectively.

Di-*t*-butyl nitroxide, as well as other di-*t*-alkyl nitroxides are unusually stable. A few other types have also been isolated, but most decay (more or less slowly) by one of several bimolecular paths. For example, if a primary or secondary alkyl substituent is present on nitrogen, a disproportionation pathway is available:

(4.2)

(MNP)

(4.3)

(DMPO)

It has been shown that both types of scavenger react with a variety of radicals, commonly at rates within three orders of magnitude of the diffusion-controlled limit (see p. 63). But each type has both advantages and disadvantages. With nitroso-compounds, the trapped radical is relatively close to the unpaired electron, and the spectrum of the adduct may display a splitting pattern which provides structural information about the trapped species. Thus a 1:3:3:1 pattern can be picked out in Fig. 4.4, which shows

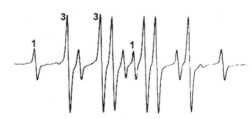

Fig. 4.4 The e.s.r. spectrum of *t*-butyl methyl nitroxide (the spin adduct of a methyl radical and MNP). The 1:3:3:1 splitting from the three methyl protons is marked. It repeats three times due to the 1:1:1 triplet arising from the [spin (I) = 1] ^{14}N nucleus.

Although some nitroso*aromatic* compounds have proved useful, nitrosobenzene itself is not particularly suitable for spin-trapping studies since the spin-adduct spectra are complicated by hyperfine splittings arising from all five phenyl-ring hydrogens.

the spectrum of the spin adduct of the methyl radical with MNP (see Eqn 4.2). This quartet splitting is not evident in the spectrum of the methyl adduct either of DMPO or of almost any of the other nitrones which have been investigated. However, biochemical applications have been concentrated on studies of oxidising radicals such as HO· and superoxide (O_2^-). The adducts of these with nitroso-compounds are either formed so slowly, or are themselves so short-lived, as to be undetectable. With nitrones, on the other hand, moderately persistent adducts are formed with both of these species, but, since splitting patterns in the resulting spectra arise only from magnetic nuclei in the trap molecule, precise comparison with reference spectra must be made. Even then, ambiguities often remain. One difficulty stems from the marked dependence of nitroxide spectra on solvent, reflecting the changing contributions of the two resonance structures to the hybrid structure as solvent polarity varies.

Another, more serious, concern, and one which has on several occasions resulted in faulty conclusions, arises from the discovery of routes that lead from spin traps to nitroxides but do *not* involve trapping of radicals. Established possibilities for nitrone traps are outlined in Eqns 4.4 and 4.5.

$$\text{(4.4)}$$

$$\text{(4.5)}$$

Despite these strictures, in competent hands the technique of spin trapping has led to important qualitative, and sometimes quantitative, information in both chemistry and biochemistry.

Nuclear magnetic resonance. One practical detail which must be borne in mind by the *n.m.r.* spectroscopist is the importance of excluding paramagnetic substances from the sample. This is to avoid line broadening effects arising from spin exchange. But there are circumstances in which this can be turned to advantage. Just as association with a rare-earth paramagnetic shift reagent can spread the chemical shifts of nuclei in a diamagnetic organic molecule, so may the unpaired electron in a stable organic radical cause large n.m.r. shifts (as well, unfortunately, as line broadening) in the expected resonance positions of nuclei in the radical, even though the direct interaction is so small that nuclear hyperfine splittings are too tiny to be resolved in the e.s.r. spectrum. A detailed discussion of this phenomenon, or of multiple resonance techniques such as Electron Nuclear DOuble Resonance (ENDOR) spectroscopy, is beyond the scope of this little text, but we shall return briefly to another n.m.r. phenomenon, dependent upon short-lived radicals as reaction intermediates, in Chapter 6.

4.2 Kinetics and thermochemistry

Rate measurements

Absolute rate measurements on gas-phase reactions of radicals date back to the 1930s, but there were serious problems with some of those early experiments. For example, NO was used as a scavenger, but various unjustified assumptions were made about the nature of its chemistry and the lack of reactivity of species derived from it. Nevertheless, much of the experimental data on the thermochemistry, and to some extent on the kinetics, of reactions of small radical species actually depends on more recent

The solvent dependence of nitroxide spectra may be interpreted in terms of the two resonance structures shown in structure **3.8** (p. 34). The more polar the solvent, the greater the contribution from the dipolar structure, hence the greater the electron-spin localisation on nitrogen, and correspondingly the greater the ^{14}N nitrogen hyperfine splitting, a_N. This dependence of a_N on solvent has been developed as a useful solvent-polarity parameter.

gas-phase work. But the emphasis in what follows will be on investigations of liquid-phase chemistry.

Relative reactivities. The earliest data on radical reactivity in liquid solution pertained only to *relative* rate measurements. We have already noted (p. 49) an approach to comparing rates of hydrogen abstraction by *t*-butoxyl from various substrates R–H. This depended on measurement of the [*t*-butanol]/[acetone] yield ratio when the radical was generated in the presence of various R–H under standard experimental conditions. In a somewhat similar fashion, the reactivities of a broad range of unsaturated molecules ('UM') towards addition of a methyl radical were investigated by determining the yields of methane and CO_2 from the thermal decomposition of acetyl peroxide in isooctane solutions containing known concentrations of UM (see Scheme 4.2). The acetyl peroxide decomposes to give methyl radicals and carbon dioxide. The methyl radicals then abstract hydrogen from the solvent forming methane. However, any competing addition of methyl to UM necessarily lowers the yield of methane, whilst that of CO_2 remains unaffected. Measurement of the $[CO_2]/[CH_4]$ ratio then permits the ratio k_{add}/k_{abstr} to be deduced. This rate ratio was called the 'methyl affinity' of UM. Some representative figures are given in Table 4.1.

Table 4.1 Relative reactivities of unsaturated species 'UM' towards addition of methyl radical, estimated by comparison with hydrogen abstraction from isooctane

'UM'	k_{add}/k_{abstr}
$MeCH_2CH=CH_2$	22
trans-$MeCH=CHMe$	7
$Me_2C=CH_2$	36
$PhCH=CH_2$	796
$CH_2=CH–CH=CH_2$	2090
Benzene	*c.* 0.4
Naphthalene	9
Anthracene	330

$$(MeCO_2)_2 \longrightarrow 2Me\cdot + 2CO_2$$

$$Me\cdot \xrightarrow[k_{abstr}]{\text{'isooctane'} \ (2,2,4\text{-trimethylpentane})} CH_4$$

$$Me\cdot \xrightarrow[k_{add}]{\text{'UM'}} (MeUM)\cdot \longrightarrow \begin{array}{c}\text{Products}\\ \text{of methylation}\end{array}$$

Scheme 4.2

In a third example, relative hydrogen-donor reactivities of R–H towards phenyl radicals were determined in a series of experiments using CCl_4 solutions of R–H, in which reaction with the CCl_4 acted as a reference. The radical precursor here was phenylazotriphenylmethane (Ph–N=N–CPh₃, see p. 48). Measurements were made of the relative yields of benzene (formed from R–H) and chlorobenzene (formed from CCl_4). The combined yields of these two products was, for each substrate, almost quantitative – a consequence of the persistent radical effect (p. 47).

A short selection of relative reactivities for hydrogen abstraction by *t*-butoxyl and by phenyl is presented in Table 4.2.

Table 4.2 Relative reactivities *per hydrogen atom* of various hydrocarbon types towards *t*-BuO· and Ph·, relative to C–H in toluene (= 1).

C–H bond in:	Abstracting radical: *t*-BuO·	Ph·
RCH_3	0.1	0.12
R_2CH_2	1.2	1.0
R_3CH	4.4	4.8
$PhCH_3$	1	1
$PhC(Me)_2H$	6.9	9.7

Absolute reaction rates. Of course, whilst compilation of data of the kind indicated above was invaluable, it fell well short of indicating the *absolute* rates of any of the radical-substrate reactions studied. However, once one *absolute* value can be provided for a member of any set of *relative* reactivities, that value provides a reference point from which to compute absolute rates for all other members of the set.

Two of the more usual methods used for these absolute rate constant measurements depend on procedures in which radicals are generated photochemically using an intense ultraviolet light source which is rapidly chopped by a rotating sector placed between the light source and the reaction vessel. In the classical rotating-sector method, the rate of termination of a chain reaction can be determined from a knowledge of the photochemical rate of initiation when the sample is exposed to light together with a measure of the average rate of consumption of one reaction partner in the chain process as a function of the speed of rotation. Without venturing into the mathematics of this, the average rate varies in a predictable way between two limiting conditions. For example, suppose there is a semicircular sector, so that the sample experiences periods of illumination and darkness of equal duration. Consider then the situation with very slow rotation. The averaged rate of consumption of reactant will be exactly half of that with constant full illumination. However, for very rapid rotation, the limiting rate can be shown to reduce only to $1/\sqrt{2}$ of that with full illumination. A full analysis of the intermediate situations then leads to a value for the termination rate constant, $2k_t$. This method was used in early studies of the reduction of alkyl bromides, RBr, by tributyltin hydride (p. 9), for which the termination is the bimolecular decay of the alkyl radicals R·.

The steady state consumption of reactant, e.g., in the above case, RBr, depends on the initiation rate, on $2k_t$, and on the rate constant k_p of the slow step in the propagation sequence, which in this case is the reaction of R· with Bu_3SnH. Therefore, once the termination rate constant has been determined, estimation of the propagation rate constant is straightforward.

The second use of a rotating sector method is in conjunction with e.s.r. spectroscopy. Signal decay after each period of illumination is very rapid, the signals are relatively weak, and special techniques are required for signal detection on a sub-millisecond time scale (necessarily short compared with signal decay). However, many signals can be accumulated into computer memory at pre-determined intervals after the illumination is chopped, and eventually a sufficiently good set of data can be obtained to achieve a satisfactory kinetic plot of radical decay during the dark period. In this way, for example, the rate constant $2k_t$ for bimolecular self-reaction (dimerisation plus disproportionation) of *t*-butyl radicals generated by photolysis of di-*t*-butyl ketone has been determined in a variety of solvents. Thus in benzene at 300 K, $2k_t = 6.5 \times 10^9 \text{ M}^{-1}\text{sec}^{-1}$. Variation of this with solvent can be related in particular to viscosity differences.

In another series of experiments, continuous ultraviolet irradiation of solutions of dicumyl peroxide ($PhCMe_2OOCMe_2Ph$) in 1,1,2-trichloro-1,2,2-trifluoroethane gives a strong spectrum of the methyl radical as the only detectable species (formed by fragmentation of cumyloxyl, p. 49). When the rotating-sector technique was applied, the second order decay was found to be contaminated by a small first-order component, presumably reflecting attack of methyl on solvent or peroxide. Trace quantities of MeCl were detected, but the yield of ethane was *c.* 90%. Analysis of the second order process gave $2k_t = 1.7 \times 10^{10} \text{ M}^{-1}\text{sec}^{-1}$ at 300 K.

Once $2k_t$ has been measured with some confidence for the methyl radical, it is in principle a simple matter to repeat the cumyl peroxide experiment with a substrate present with which methyl reacts, to monitor the resulting increase in the first order component of the disappearance of methyl, and thence to deduce the absolute rate of reaction of methyl with the added substrate. This approach has permitted the rate constants for addition of methyl to several unsaturated molecules (UM, see above) to be determined. The relative values of the rate constants obtained in this way agreed well with the relative methyl affinities for the same compounds determined in the older work. This was considered to justify estimating absolute rate constants for methyl radical addition to well over 200 different substrates, by relating the extensive published set of methyl affinities to the small set of newly measured rate constants.

Variants of the basic kinetic e.s.r. procedure have replaced the rotating sector with intense nanosecond (or faster) repeating flashes of light from a pulsed laser, or by very short (as little as 10^{-11} sec) bursts of fast electrons from a small linear accelerator ('pulse radiolysis'), and have used kinetic *optical* spectroscopy for radical detection. These methods are unsuitable for the direct study of simple alkyl radicals, since no convenient absorption features are present. They have, however, been extensively used where visible or ultraviolet monitoring is possible, and highly automated systems have been installed in numerous research facilities. The radiolytic techniques have been particularly applicable to aqueous systems, whilst flash photolytic methods have more commonly investigated reactions in organic solvents. Both of these methodologies have also been used in gas-phase studies.

One approach to measurement of rate data for reaction of radicals which have no suitable absorption features is to carry out experiments with 1,1-diphenylethene (**4.5**) *mixed* with the substrate 'X' of interest. Monitoring the rate of growth of the benzhydryl (**4.6**) spectrum in two sets of experiments in which first the concentration of **4.5** is varied and then that of X is varied allows both $k_{4.5}$ and k_X to be determined.

$$R\cdot + Ph_2C=CH_2 \xrightarrow{k_{4.5}} Ph_2\overset{\cdot}{C}-CH_2R$$
$$\textbf{4.5} \qquad\qquad\qquad \textbf{4.6}$$

$$R\cdot + X \xrightarrow{k_X} [RX]\cdot$$

Free-radical clocks. We have seen above how the determination of one or more absolute rate constants may be used in conjunction with accumulated *relative* rate data to estimate absolute reaction rates for a wide variety of reactions. When the reference reaction is a unimolecular rearrangement, it has been designated a 'clock' reaction or 'free-radical clock'. A typical example is the hexenyl → cyclopentylmethyl rearrangement. It is of course first necessary to measure the rate of the clock reaction, and in the case of the hexenyl rearrangement this was originally accomplished by means of the tributyltin hydride reduction of 6-bromo-1-hexene. This gives a mixture of 1-hexene and methylcyclopentane in proportions determined by the competition between unimolecular rearrangement (k_R) and bimolecular reaction with the tin hydride (k_2), as indicated in Scheme 4.3. Provided that the tin hydride is in large excess, so that its concentration remains effectively

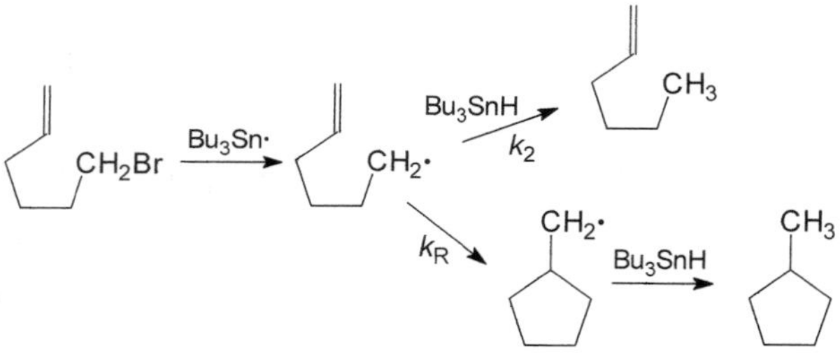

Scheme 4.3

constant, the ratio of hydrocarbon products will be given by Eqn 4.6. Since a reasonable assumption would be that k_2 is essentially the same as the value measured for e.g. n-hexyl, straightforward experimental measurements allow k_R to be determined (*c.* 1.5×10^5 sec^{-1} at 65° C – subsequently redetermined

$$k_R/k_2 = \frac{[\text{methylcyclopentane}]}{[\text{Bu}_3\text{SnH}][\text{1-hexene}]} \qquad (4.6)$$

by direct methods). Exactly the reverse of this procedure allows reactions of primary alkyl radicals with various substrates to be 'timed'. For example, one of the first estimates of a rate constant for a spin-trapping reaction (see above) was obtained by monitoring the growth of hexenyl and cyclopentylmethyl nitroxides formed by trapping with MNP (Scheme 4.4). The measured rate constant, $k_{2(ST)}$, was *c.* 9×10^6 sec^{-1} at 40° C, which is characteristic of the efficiency of many spin-trapping reactions.

An interesting problem was encountered in the experiments outlined in Scheme 4.4. The two spin adducts, both primary alkyl nitroxides, had very similar spectra which were almost exactly superimposed. To complete the experiment, this difficulty was overcome by incorporating ^{13}C at C(7) in the initial peroxide. The additional ^{13}C-splitting in the spectrum of the cyclopentylmethyl nitroxide satisfactorily avoided the overlap difficulty. The isotopically labelled carbon is starred in Scheme 4.4.

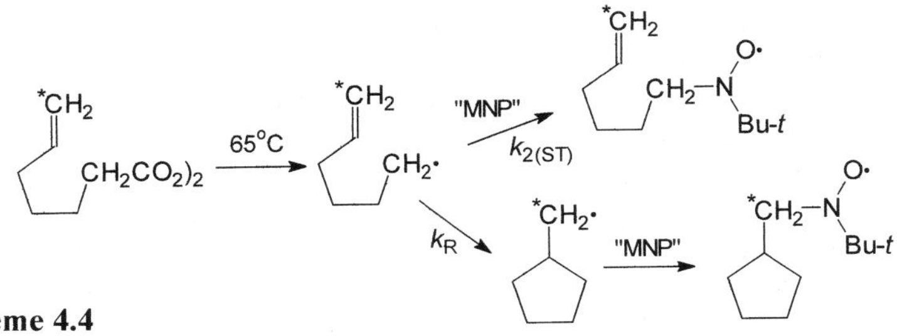

Scheme 4.4

The hexenyl clock is appropriate for a range reactions of primary alkyl-radicals, but when the bimolecular process whose rate is to be measured is very much faster or very much slower than the above spin-trapping example, unimolecular reactions with higher or lower rate constants are more appropriate. Rates of rearrangement of a small selection of free-radical clocks are collected in Table 4.3.

Table 4.3 First order rate constants for some unimolecular rearrangements of primary alkyl radicals which have been used as free-radical clocks.

Reaction	Conditions	k (sec^{-1})
$\text{PhCMe}_2\text{CH}_2\cdot \rightarrow \cdot\text{CMe}_2\text{CH}_2\text{Ph}$	benzene; 298 K	7×10^2
Hexenyl → Cyclopentylmethyl	benzene; 338 K	1.5×10^5
	C_6F_6; 303 K	2.1×10^8
	benzene; 303 K	3×10^{11}

Thermochemistry of radicals and bond dissociation energies

Although we have, in Chapter 3, emphasised the relationship between bond-energy data and kinetics, it is likely that for the organic chemist an awareness of rate data will be more important than one of reaction energetics, since it is kinetic factors that determine the success or failure of one possible reaction pathway in competition with another.

In fact, the experimental approach to the C–H bond dissociation energy of e.g. methane involves the experimental methods of the gas kineticist. Typically, the activation barriers for both the forward and back reactions of Eqn 4.7 are measured separately. The difference between these gives the heat of reaction. Now the heat of reaction 4.7 is evidently the difference between the CH_3–H bond dissociation energy, and that of H–Br. Since the latter is known from its gas-phase heat of formation, the former may be estimated.

$$CH_3\text{---}H + Br\cdot \rightleftharpoons CH_3\cdot + H\text{---}Br \qquad (4.7)$$

Measurements of the kinetics of forward and back reactions provide a value for the equilibrium constant, whence ΔG may be determined. Other experimental approaches to radical thermochemistry have attempted direct equilibrium measurements. For example, the exchange of iodine between alkyl radicals in solution is essentially diffusion controlled, so that, in an e.s.r. experiment in which radicals are generated by abstraction of iodine from mixtures of RI and R'I, it has been possible to monitor the concentrations of R· and R'·, and hence to estimate an equilibrium constant for Eqn. 4.8.

$$R\cdot + R'\text{---}I \rightleftharpoons R\text{---}I + R'\cdot \qquad (4.8)$$

Despite a range of experimental approaches, even now error limits of between ± 2 and ± 6 kJ mol^{-1} are put on the strengths of C–H bonds in simple alkanes, reflecting uncertainties and assumptions in the various experimental procedures.

For weaker bonds, such as S–H, O–H in phenols, or C–H in skipped dienes, alternative experimental methods have been developed. However, we have today reached the point at which molecular orbital theory will usually give quantitative bond-dissociation energy data of an accuracy comparable to that of experiment. For small radicals this may involve *ab initio* procedures. For larger species, data can be deduced by analogy with suitable small model systems, or, and often with quite acceptable precision, by means of rapid, semi-empirical routines on a personal computer.

5   Autoxidation – a case study

Inevitably, in a short introductory text, topic coverage has to be selective. One ubiquitous radical process is 'autoxidation' – oxidation by atmospheric oxygen. We shall now look at aspects of autoxidation in a little more detail, paying attention to the energetics of the individual steps, to certain complexities which are important with particular substrates, and to some of the ramifications of autoxidation, especially in biological systems.

Propagation. The basic chain-propagating steps for autoxidation were introduced on p. 9, and are repeated here in Scheme 5.1. The slow step is the abstraction of hydrogen from the substrate (Eqn 5.2). The rate of combination of substrate-derived radical R· with oxygen (Eqn 5.1) is usually close to the diffusion-controlled limit.

For a wide variety of carbon-centred radicals, and in solvents ranging from water to cyclohexane, k_1 has been measured, using pulse-radiolysis techniques, to be greater than $10^9 \ \mathrm{M^{-1} \ sec^{-1}}$ at 298 K.

$$R\cdot \ + \ O_2 \ \xrightarrow{k_1} \ RO_2\cdot \tag{5.1}$$

$$RO_2\cdot \ + \ R{-}H \ \xrightarrow{k_2} \ RO_2H \ + \ R\cdot \tag{5.2}$$

Scheme 5.1

It should therefore be evident that the ease of autoxidation of a given substrate depends markedly on the strength of its weakest C–H bond. In addition, since the unpaired electron in a peroxyl radical is localised on oxygen, the radical is necessarily electrophilic. One manifestation of this polar character is in the relatively facile autoxidation of ethers, where the hazards associated with peroxide formation are well known. For diethyl ether, the hydrogen abstraction step (Eqn 5.3) is endothermic by *c.* 25 kJ mol^{-1}, but the activation barrier is lowered by dipolar stabilisation of the transition state (**5.1**) (to *c.* 40 kJ mol^{-1}).

Particularly notorious in the literature of laboratory accidents is the peroxide which is formed from diisopropyl ether. The solid which may separate from solution is highly shock-sensitive. That solid is not, however, the initially formed hydroperoxide, but rather a macrocyclic product (**5.2**) derived from it.

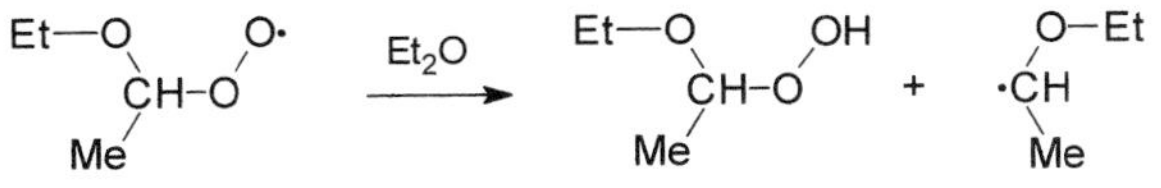

$$\tag{5.3}$$

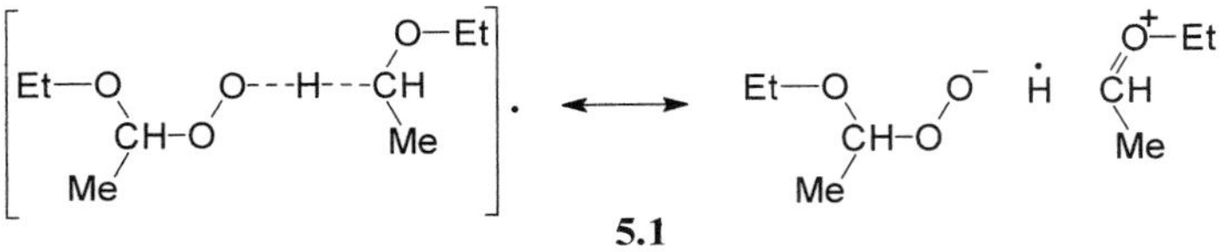

5.1

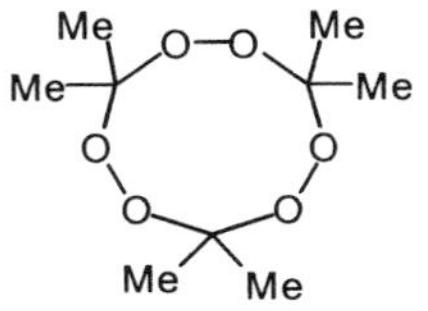

5.2

In the case of hydrocarbons, the durability of commercial products such as lubricating oils might be enhanced by minimisation of the number of relatively weak *tertiary* C–H bonds in the molecular structure. A patent was registered, many years ago, for the incorporation of deuterium into the tertiary sites in fine lubricating oils. The object was to utilise the kinetic isotope effect (where C–D is effectively stronger than C–H) in order to prolong the service interval for superior quality mechanical clocks and watches.

A variety of experimental techniques has been brought to bear on measurement of propagation rate constants (k_2), and some representative values are given in Table 5.1. Of special significance are the relatively high

Table 5.1 Approximate rate constants (303 K) for various reactions ROO· + RH → ROOH + R· (per reactive hydrogen at 303 K).

R–	k_2 $(M^{-1} sec^{-1})$
EtMe$_2$C–	7×10^{-3}
PhCH$_2$–	8×10^{-2}
RCH$_2$(CH=CH)CHR'–	0.5
(RCH=CH)$_2$CH–*	31
PhC(=O)–	3.3×10^4

*The isomeric peroxyl R(RCH=CH–CH=CH)CHOO· is involved; see text.

Glycerides are the esters of glycerol, a trihydric alcohol. Naturally occurring triglycerides include fats and oils from animal and vegetable sources. In these the glycerol is fully esterified by mixtures of long-chain (commonly C$_{18}$) fatty acids, important examples of which are listed below. Fats differ from oils solely in melting behaviour, but usually have a higher proportion of saturated fatty acid residues. The phosphoglycerides, represented here by phosphatidyl choline (**5.3**), incorporate a phosphate residue.

glycerol   a triglyceride

C$_{18}$ fatty acids (RCO$_2$H):

Stearic, R = C$_{17}$H$_{35}$–

Oleic, R =

Linoleic, R =

Linolenic, R =

5.3

values encountered with structures incorporating doubly allylic hydrogen – the so-called 'skipped' dienes (note also entry 6 in Table 3.1, p. 32). This structure is commonly found in triglyceride vegetable oils, as well as in the closely related phosphoglycerides, *inter alia* important constituents of biological cell membranes.

The seeds of certain plant species, including a variety of common flax (linseed) and walnut, produce oils, the molecules of which are particularly rich in skipped diene features. If thin films of such oils are left exposed to the air, they will slowly solidify. These oils are therefore designated 'drying oils'. This drying characteristic led to their use as liquid media in which to suspend finely powdered inorganic pigments to create the paints which were applied to board or canvas in the construction of the now centuries old pictures which survive in our historic art collections. Many of the features of modern paint chemistry derive from these ancient techniques.

The drying phenomenon involves autoxidation, but is in fact exceedingly complex. The initial hydrogen abstraction forms a stabilised pentadienyl radical which combines with oxygen at a terminal carbon. The resulting peroxyl radical abstracts hydrogen from a second skipped diene to give a hydroperoxide that now incorporates a *conjugated* diene system (Scheme 5.2). The hydroperoxide formed in this initial reaction of the oil molecule is unlikely to solidify. Solidification is the result of further reactions which yield a cross-linked polymer network.

With a simple, non-terminal alkene, peroxyl radicals will react by two competing processes, namely addition, and abstraction of allylic hydrogen. With a skipped diene, however, the high reactivity of the doubly allylic hydrogen markedly favours abstraction. But bringing the two double bonds into conjugation raises the rate of addition so that it is now competitive with abstraction from the skipped diene. The adduct from the conjugated diene is a resonance-stabilised allylic radical which can react further with oxygen to create a new peroxyl (e.g. as in Scheme 5.3).

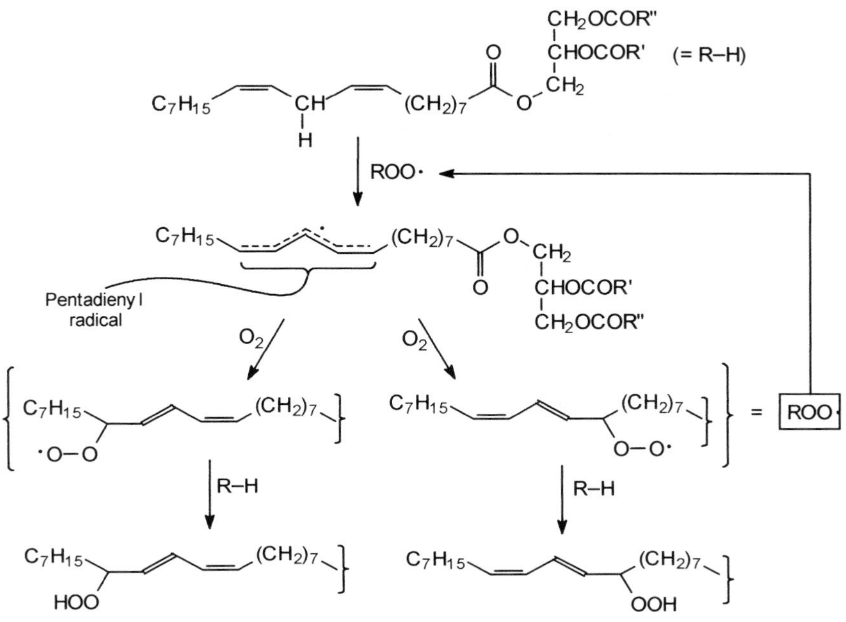

Scheme 5.2

Since the original triglyceride had three fatty acid 'tails', two, or all three of which were likely to contain at least one skipped diene unit, there is the possibility that peroxyl addition may be either intramolecular or inter-molecular. The intermolecular process is the beginning of a polymerisation sequence. Furthermore, because more than one 'tail' is likely to contain unsaturation, then as oxidation proceeds cross-linking will occur to form a three-dimensional polymeric network, eventually yielding a rigidly solid film.

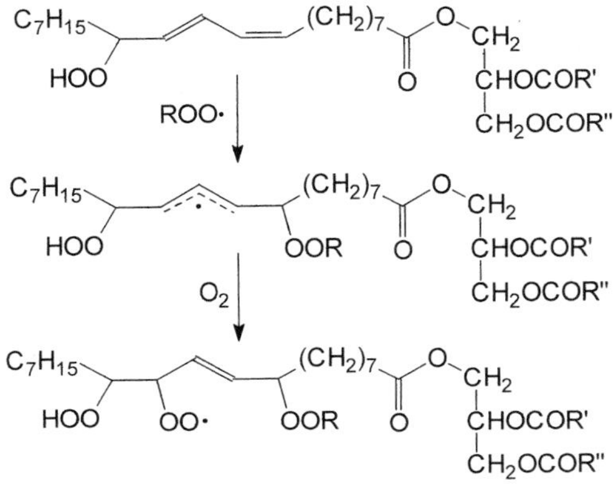

Scheme 5.3

This is still only part of the story. The absorption of oxygen that is implicit in the above reactions is readily detectable by the substantial weight increase which occurs. In fresh linseed oil, this may be as great as 40% after

1-2 days. Thereafter, a slow weight decrease sets in. If volatile materials are collected, and are then analysed by vapour phase chromatography, a multitude of low molecular weight organic compounds can be identified. A principal route to these involves formation of an alkoxyl radical (either by induced decomposition of hydroperoxide – or possibly by a termination reaction, see below), which then fragments. An illustrative sequence, in which cobalt(II) induces peroxide decomposition, and octanal is produced, is displayed in Scheme 5.4. Some of these volatile fragments have powerful and unpleasant odours. Together with evaporating solvent, they contribute to the smells associated with the drying of oil-based household paints. Interestingly, very similar chemistry contributes to the 'off-flavours' associated with foodstuffs that have deteriorated on storage, although in these cases the proportion of polyunsaturated fatty acid tails in the glyceride components will usually be much lower than that in the drying oils.

Organic soluble cobalt salts are commonly present in paint formulations as 'driers' (see p. 22). Note also that dialkyl peroxides are far more resistant than are hydroperoxides or H_2O_2 towards this type of induced decomposition.

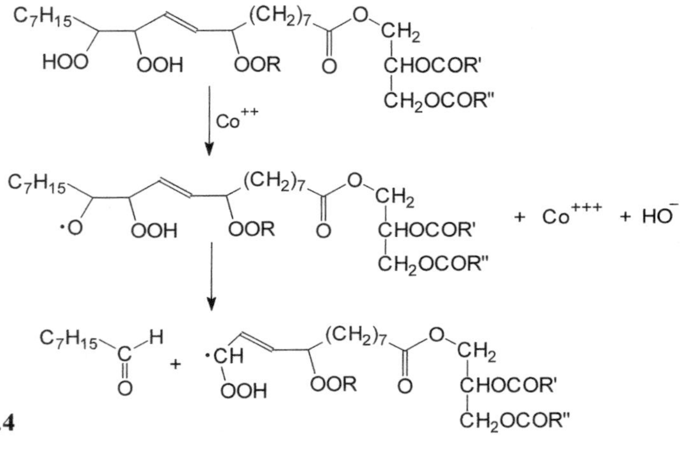

Scheme 5.4

Termination. As in any chain reaction, propagation steps in autoxidation must compete with termination. Whilst several possible radical-radical processes may be written, by far the most important in autoxidation reactions is the interaction of two peroxyl radicals (exceptions may occur in the presence of very low oxygen concentrations). Interaction of two peroxyl radicals is much slower than the diffusion limit (i.e. these radicals are relatively persistent), so that their concentrations build to levels that are detectable by some of the more sensitive spectroscopic methods, such as e.s.r. The range of rate constants at 303 K for termination is quite wide (Table 5.2), with that for tertiary peroxyls being by far the lowest. This reflects a change of mechanism should there be no α-hydrogen present. For the tertiary peroxyls, reversible dimerisation to tetroxide is accompanied by decomposition to oxygen and two alkoxyl radicals (Eqn 5.4). Cage recombination of the latter gives di-*t*-alkyl peroxide; the alkoxyl radicals which escape the cage can continue the autoxidative chain by reacting with further molecules of substrate. For primary and secondary peroxyls, the so-called Russell mechanism (Eqn 5.5) is generally believed to be operative, although alternative proposals have from time to time been put forward to rationalise the much faster decay of these species. Support for the Russell

$$2ROO\cdot \longrightarrow \qquad \longrightarrow \qquad \mathbf{5.4} \qquad (5.5)$$

(R = *primary* or *secondary* alkyl)

mechanism includes the observation that singlet oxygen (represented as **5.4**; see p. 83) is produced, as might be expected for a six-electron pericyclic process. However, contrary evidence includes the observation that the ratio of alcohol to ketone rapidly drops to values much less than unity as the temperature of the reaction is lowered below ambient.

Table 5.2 Approximate rate constants for self-reaction of peroxyl radicals (2ROO· → products) in hydrocarbon solvents at 303 K.

R in ROO·	$2k_t$ (M^{-1} sec^{-1})
Et	4×10^7
Cyclohexyl	2×10^6
Ph_2CH	3×10^7
t-Bu	1.3×10^3
$PhCMe_2$	6×10^3

Initiation. Most physical studies of peroxyl-radical formation and of reactions of peroxyls depend on active initiation, typically by thermolysis or, more usually, photolysis of azo compounds such as AIBN (p. 16). Other appropriate photochemical procedures have also been used, including ones depending on abstraction of hydroxyl hydrogen from hydroperoxides.

Photoinitiation of autoxidation is likely to be accentuated by the photolysis of the initially formed hydroperoxides. Two simple but quite distinct examples, depicted overleaf, involve benzaldehyde and cyclohexene. Oxidation of benzaldehyde, which gives peroxybenzoic acid, occurs readily with the pure compound, since photo-excitation of the carbonyl group generates a species, conveniently represented as the biradical (**5.5**) (see p. 82), which abstracts hydrogen from a second molecule of aldehyde. With cyclohexene, photo-oxidation depends on absorption of light energy by a sensitizer ('S') which transfers its excitation energy to molecular oxygen. The resulting singlet oxygen molecule (**5.4**) is a powerful enophile and forms cyclohexene hydroperoxide (**5.6**) in a pericyclic process. Unless the sensitiser is excited by light of a wavelength which is not absorbed by the peroxide, further photolysis will produce radicals.

Initiation:

Propagation:

Photoinitiated autoxidation of benzaldehyde

Photosensitised formation of, and photolysis of cyclohexene hydroperoxide

The spontaneous initiation of autoxidation, on the other hand, is more problematic. In some circumstances, initiation can be traced to the presence of impurities. Pre-existing hydroperoxide impurities, for example, may be induced to decompose by the presence of traces of transition-metal ions. Alternatively, other modes of molecule-induced homolysis may be important, as in the spontaneous formation of radicals from styrene (see pages 17-19). It seems possible that, in some cases at least, this may involve a direct bimolecular reaction between R–H and O_2.

Antioxidants. Although autoxidation may be a desirable phenomenon in the paints and coatings industry, it is clearly something to be avoided in many other circumstances. To this end, many commercial products, e.g. rubber, lubricating oils, and numerous foodstuffs, contain added 'antioxidants'.

Bear in mind that one initiating radical may cause the oxidation of hundreds or even thousands of substrate molecules. If an additive can interrupt the chain-propagating sequence, degradation of the substrate can be dramatically slowed. Particularly effective in achieving this are certain sterically hindered phenols, notable among which are 2,6-di-*t*-butyl-*p*-cresol ('butylated hydroxytoluene', BHT; **5.7**) and 2-*t*-butyl-4-methoxyphenol (butylated hydroxyanisole, BHA; **5.8**). These 'chain-breaking antioxidants' react with peroxyl radicals by hydrogen-atom transfer from oxygen to form resonance-stabilised phenoxyls. The latter are then insufficiently reactive to propagate the autoxidative chain by abstracting hydrogen from substrate. Instead, they may rapidly intercept a second peroxyl, yielding non-radical product (e.g. **5.9**). In the example of the hydrocarbon tetralin (**5.10**), the

substrate-derived peroxyl radical (**5.11**) reacts with BHT some 1,000 times faster than with the hydrocarbon, so that millimolar concentrations of BHT will be sufficient to suppress autoxidative chains that consume more than a very few molecules of substrate.

A variety of other materials have been employed in order to inhibit autoxidation, and several different mechanisms are involved. For example, certain compounds may be added which minimise photoinitiation by efficiently absorbing ultraviolet light and then dissipating excitation energy by processes which do not induce radical production. A more detailed discussion is unfortunately beyond the scope of this introductory text.

Polyunsaturated fatty acids revisited: autoxidation in biology. It was first suggested some 50 years ago that radicals might be important in human disease, especially in cancer, and involvement of radicals in a wide range of biochemical mechanisms has been vigorously investigated during the past three decades. Once again, space limitations allow us no more than super-ficial coverage, which will be focused on aspects of autoxidation. This has been perceived as a particularly important area, because of the presence of easily oxidised lipids throughout the body. In particular, those phospho-glyceride molecules which incorporate polyunsaturated fatty acids containing skipped-diene units, have been the subject of intensive scrutiny. In order to simplify the experimental investigation of these it has been common practice to concentrate on the methyl ester of linoleic acid (**5.12**). The naturally

5.12 Methyl linoleate

occurring linoleic acid has both double bonds *cis*, and the ester is relatively easily obtained in sufficiently pure form for detailed study. Some of the chemistry which has been thus disclosed is of more general significance.

It was noted above that the pentadienyl radical from a skipped diene reacts with oxygen at the terminal carbon atom. In accord with this, methyl linoleate actually forms *four* hydroperoxides (**5.13–5.16** – as racemates), including two *trans,trans* isomers (structural types omitted from Scheme 5.2). This is at first sight surprising, since pentadienyl radicals have been

5.13   **5.14**

5.15   **5.16**

shown (by e.s.r. spectroscopic studies) to be configurationally stable on the millisecond time-scale of their existence during the autoxidation sequence. How, then, can the *trans,trans* hydroperoxide isomers arise, assuming that the pentadienyl radical has the configuration shown (**5.17**)? (This structure both retains the *cis* geometry of the 1,2 and 4,5 bonds, and avoids steric congestion by adopting *trans* geometry about the 2,3 and 3,4 bonds).

5.17

A further subtlety arises with oxidation of (monounsaturated) oleic acid. This is, of course, slower than that of the skipped dienes. When one positional isomer of the derived hydroperoxide is exposed to a radical initiator, rearrangement also occurs , this time by 1,3 (i.e. allylic) migration of OOH. In this case, however, there is very little exchange with external O_2. Selective labelling of only one hydroperoxide oxygen, shows that this is scrambled in both the isomeric, *and* in the unrearranged hydroperoxide. A mechanism involving dissociation and predominant recombination within the solvent cage has been proposed.

Contributing to the evidence which led to a solution of this problem were the discoveries: (1) that higher linoleate concentrations give higher proportions of *cis,trans* products; (2) that a single *cis,trans* isomer of hydroperoxide could be transformed into a mixture of all four isomers by the action of a free-radical initiator, and that during this process the peroxide oxygens exchanged with isotopically-labelled oxygen gas admitted to the reaction system.

The interpretation depends on a reaction which we have as yet overlooked in our discussion of autoxidation, namely the *reverse* of radical combination with molecular oxygen. This is evident only in highly stabilised carbon radicals, such as triarylmethyls and, as here, pentadienyls. Under conditions of kinetic control, *cis,trans* peroxyls will lead on to form *cis,trans* hydroperoxides, but if the peroxyls can revert to dienyl radicals they may do so from a conformation different from that in which they were formed, giving a dienyl radical stereoisomeric with **5.17**, e.g. **5.18**. A possible fate of this species is combination with oxygen to give a *trans,trans* peroxyl and hence a *trans,trans* hydroperoxide. In the limit, thermodynamic control would yield an equilibrium mixture of the four hydroperoxide isomers.

Amongst the biological consequences of autoxidation is the development of deposits of arterial plaque as in the atherosclerosis of coronary heart disease; advice on the minimisation of these and related problems includes a reasonable dietary intake of the antioxidant vitamins, especially C and E. Vitamin E (α-tocopherol; **5.19**), nature's principal fat-soluble antioxidant, is a phenol which behaves in a similar fashion to the synthetic BHT and BHA mentioned above. When present during the model studies with methyl linoleate, α-tocopherol has the expected effect of lowering the rate of autoxidation. It also has the effect of dramatically reducing the quantities of the *trans,trans* hydroperoxides which are formed. The explanation for this is that the tocopherol transfers hydrogen to peroxyl radicals with a rate constant of *c.* 3×10^6 M^{-1} sec^{-1} at 300 K, thus allowing little opportunity for the peroxyl intermediates to revert to dienyl radicals.

These observations, which are evidently pertinent to lipid autoxidation in biological systems, provide a level of understanding far superior to that of many early biochemical investigations, in which the extent of oxidation was frequently monitored only by rather crude colorimetric procedures.

Two other features of biochemical antioxidant behaviour are of interest. Firstly, vitamin C (ascorbic acid, **5.20**) is a water soluble antioxidant, but

As another example, there is increasing evidence that the progress of many of the neurodegenerative diseases of old age may be slowed by administration of these antioxidant vitamins.

there is an interfacial synergy with vitamin E, whereby the tocopheryloxyl radical may be reduced back to α-tocopherol.

The second point involves a physicochemical subtlety. It is generally considered that the phenoxyl derived from a phenolic antioxidant will not propagate autoxidation. However, there is a particular circumstance in which this is not the case. This arises when only a single autoxidative chain is propagating within a single particle of a hydrophobic mesophase such as a micelle or liposome (and without a high vitamin C concentration in the aqueous phase). Reaction with a phenol, e.g. vitamin E, under these conditions affords a phenoxyl radical which will have a dramatically extended lifetime. This is because normal diffusive processes are suppressed, and the frequency of radical-radical encounters, which depends on one radical transferring between particles, is greatly diminished. The result is that very slow hydrogen abstraction from lipid by the phenoxyl radical may now occur, such that the phenol takes on the unfamiliar role of 'pro-oxidant'. This behaviour is exactly similar to the formation of very high molecular weight polymers in the industrially important process of emulsion polymerisation, in which only a single radical chain is growing in any one emulsion droplet. It is not yet clear, however, whether this phenomenon does in fact mediate any biological oxidations, such as, for example, arterial plaque formation.

Arachidonic acid oxidation. Not all autoxidative processes in biology are uncontrolled. An especially important substrate for oxidation is arachidonic acid (**5.21**), a C_{20} tetraene fatty acid. This is the biochemical precursor to the prostaglandin hormones, as well as to a number of other biochemically important molecules, including the leukotrienes (important in the immune response), prostacyclin, and the thromboxanes. All of these are derived, at least in part, by enzymatically controlled autoxidations. We shall consider here only the route to the prostaglandins, controlled by a 'cyclooxygenase' enzyme (Scheme 5.5). Initial hydrogen abstraction from C(13) is followed by reaction with O_2 at C(11). The resulting peroxyl radical has the character of a 1,2-dioxa-5-hexenyl radical (see inset) and, like other hexenyls (p. 13), it cyclises. The derived cyclic peroxide again has hexenyl character, and a second intramolecular addition ensues. This time the result is an allylic radical which reacts with a second molecule of oxygen at what was originally C(15), ultimately yielding the hydroperoxide **5.22**, known as prostaglandin G_2. Further metabolic chemistry leads to an array of related structures.

It is interesting that weak prostaglandin-like biological activity is detectable in the crude reaction mixture arising from an *in vitro* oxidation of arachidonic acid. That enzyme mechanisms are operating in the biological system is immediately evident, however, from the strict stereochemical control which gives only the product shown.

Triphenylmethyl revisited. Prominent in our historical introduction (p. 1) was a note of Gomberg's pioneering discovery of triphenylmethyl ('trityl') radicals. When, in his initial experiments with chlorotriphenylmethane and mercury, air was *not* excluded from the reaction, a product was isolated which contained oxygen. Gomberg correctly identified this as trityl peroxide (**5.23**), and, after careful consideration, he proposed that the trityl radical dimer had the (sterically very congested) structure **5.24**. It was more than sixty years before this was shown to be in error. Assignment of the correct structure, **5.25**, in which coupling has occurred between the methyl carbon of one trityl radical, and the *para* position of a second, at last explained some apparently anomalous results related to the formation of the trityl peroxide.

The simple, even 'obvious', interpretation of peroxide formation is that a trityl radical reacts with oxygen, and that the resulting peroxyl ($Ph_3COO\cdot$) is then intercepted by a second trityl. But there is a problem. There was kinetic evidence for the operation of a chain reaction in which the peroxyl reacts with the dimer. For this to occur with **5.24** would necessitate the occurrence of the rare (p. 10) S_H2 substitution at saturated carbon. This anomaly is neatly avoided with the new formulation for the dimer – addition to the exocyclic double bond being followed by fragmentation, as shown below.

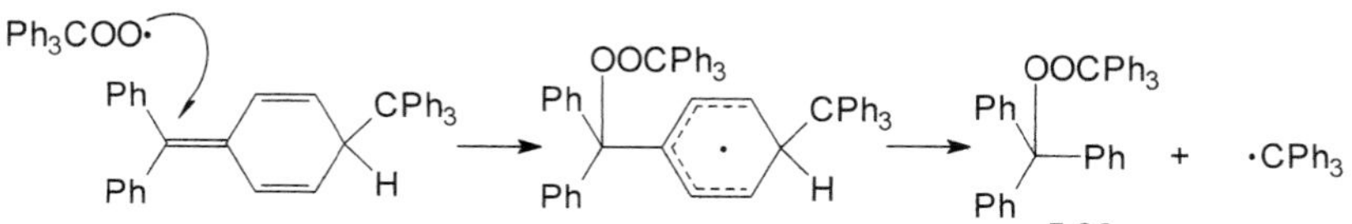

$$Ph_3C—CPh_3$$
5.24

Not only would **5.24** be a highly strained structure. The congestion in triphenylmethyl itself is such that each benzene ring is skewed out of plane, rather like the blades of a propeller, in order to minimise the steric interference between the *ortho* hydrogens.

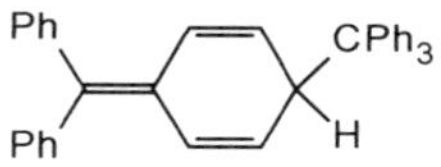

5.25

This short excursion into autoxidation, which has focused exclusively on condensed phase behaviour, has attempted to provide a somewhat deeper insight into one particular class of radical reactions. It is, of course, a class that has ramifications throughout many branches of food and manufacturing science, as well as important consequences for the health – or otherwise – of the living organism. As such, our discussion has perhaps gone a little way towards forging a link between the academic study of radical chemistry and the world outside!

The complexities of gas-phase autoxidation are extensively documented. In combustion chemistry above c. 700-800 K, even the simple alkylperoxyls are unstable with respect to R· and O_2, and other pathways open up for the reactions between these species.

6 Radical ions, radical pairs, and biradicals

Arguably, a basic knowledge of radical chemistry provides a valuable, if not absolutely essential, introduction to key aspects of several other important topics, including organic photochemistry, the chemistry of carbenes and nitrenes, and many examples of oxidation and reduction. The unifying theme, at least in the organic chemist's perception of bonding based on a molecular orbital framework, is the single-electron occupancy of an individual spatial orbital.

6.1 Radical ions

Students of organic chemistry will most probably have encountered radical ions for the first time when learning of some reduction reactions of organic carbonyl compounds: representative of these are the the pinacol reduction of ketones, depicted in Scheme 6.1, and the acyloin reaction of esters. And in the laboratory, they may have seen the blue colour of benzophenone ketyl (**6.1**), indicative of the removal of last traces of moisture and molecular oxygen from ether solvents dried by refluxing over metallic sodium in the presence of the ketone.

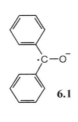

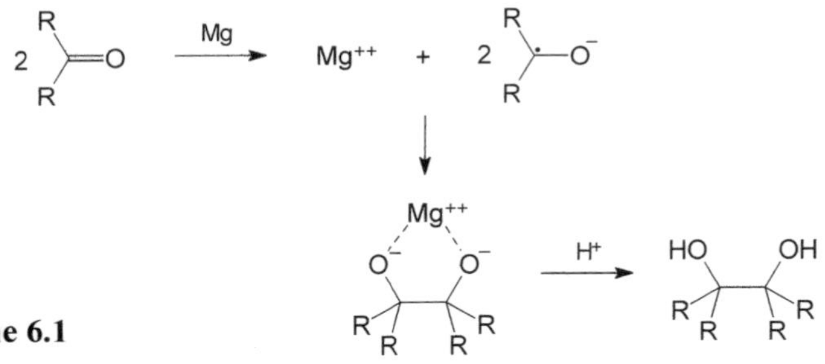

6.1

Scheme 6.1

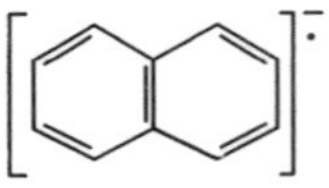

In the early days of electron spin resonance spectroscopy, such rigorously dried and oxygen-free solvents were employed for the detection and spectroscopic characterisation of radical ions formed from a variety of organic molecules. Thus naphthalene radical anion (**6.2**) may be generated by reduction of naphthalene in tetrahydrofuran, either chemically, e.g. by potassium metal, or electrochemically. Under these specially controlled conditions, modest concentrations of radical anions derived from highly delocalised molecules can be maintained indefinitely. However, like electrically neutral radicals, such species are more familiar in organic chemistry as reactive intermediates. The same is true for radical cations, formed during certain oxidation processes. Crucial to both types of reaction is single-electron transfer (SET).

6.2

Under conditions where it does not persist, naphthalene radical anion is a key intermediate in, for example, the Birch reduction of naphthalene (Scheme 6.2). Sodium dissolves in liquid ammonia as Na^+ and a solvated electron, $(e^-)_s$. The solvated electron transfers to dissolved naphthalene to form the radical anion, but, under Birch conditions, there is normally an alcohol present. The naphthalene radical anion behaves as a base, and is protonated to give the intermediate benzocyclohexadienyl radical (**6.3**). This is rapidly reduced further to the corresponding anion. A second protonation then yields 1,4-dihydronaphthalene (**6.4**). Although an isolated benzene ring may also be reduced under Birch conditions, this is slower, and **6.4** is

With larger quantities of sodium, the second aromatic ring may also be reduced, to give 1,4,5,8-tetrahydronaphthalene (see also p. 21):

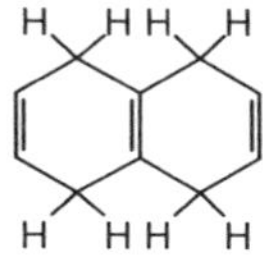

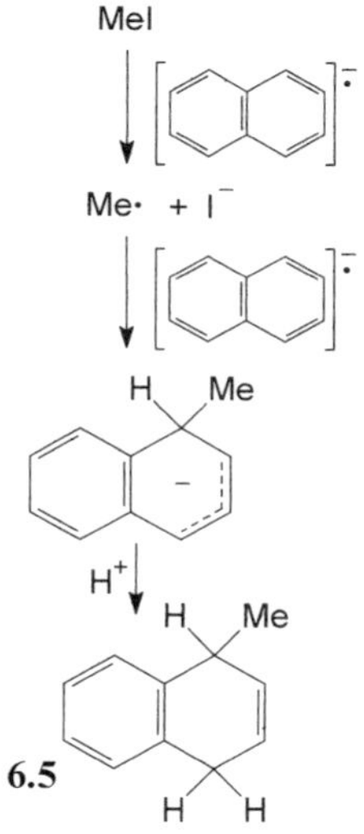

Scheme 6.2

readily isolated if the correct amount of sodium has been used. In a variant of the reaction, methyl bromide or iodide may be added, which leads to the 1-methyl-1,4-dihydronaphthalene (**6.5**). However, the picture here is complicated. A high yield is achieved under conditions in which no alcohol is present, and in these circumstances a second electron is thought to be transferred to naphthalene to give the dianion; it is this which is believed to be alkylated. In contrast, experiments using solutions of authentic radical anion in dimethoxyethane do give some **6.5**, but in poor yield, together with an array of additional products. In this case it may at first appear that the radical anion is behaving as a nucleophile, but there is evidence to suggest that this should properly be interpreted in terms of SET between the radical anion and methyl halide to generate a methyl radical. The radical is then intercepted by a second naphthalene radical anion in what amounts to a radical-radical ion coupling process. This sequence is set out in Scheme 6.3.

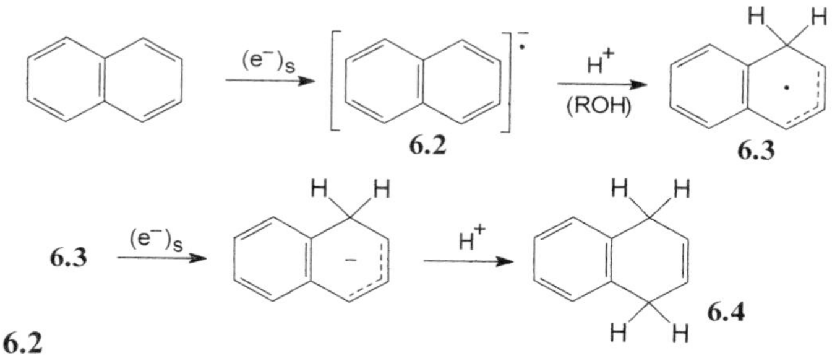

In the naphthalene radical anion the unpaired electron occupies the lowest energy π^*-antibonding orbital, and is therefore delocalised over the entire carbon framework. This is reflected in the e.s.r. spectrum (see Fig. 6.1),

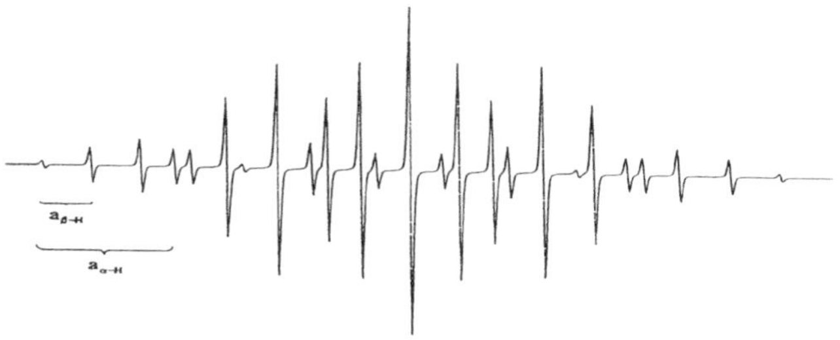

Scheme 6.3

Fig. 6.1 Electron spin resonance spectrum of the naphthalene radical anion.

the twenty-five lines of which may be interpreted as a quintet of quintets, consistent with interaction with a set of four equivalent α-protons, and a (smaller) interaction with a set of four β-protons.

The naphthalene radical cation is structurally rather similar, except that in this case the single electron occupies the highest energy π-bonding orbital. We saw on p. 22 how this species is important in the electrochemical acetoxylation of naphthalene. A somewhat similar reaction occurs when naphthalene is oxidised by manganese(III) acetate. Many other familiar inorganic reagents behave as one-electron oxidising agents, including $Fe(CN)_6{}^{3-}$, lead tetraacetate, and potassium persulphate ($K_2S_2O_8$).

Noted above was the one-electron reduction of methyl iodide to give a methyl radical in a process which is generally termed 'dissociative electron capture'. Extensive studies of a variety of systems have focused on the concertedness or otherwise of the electron capture and departure of the negatively charged leaving group. We shall skate round this subtlety, and merely note that dissociative electron capture is believed to be an important step in a number of substitution reactions which have the mechanistic designation $S_{RN}1$ (substitution radical nucleophilic, with unimolecular dissociation of the radical ion intermediate). An example is given in Scheme 6.4. This shows the successful displacement of aromatic bromine by an enolate anion. The reaction is effected in liquid ammonia, is accelerated by light, and is inhibited by radical traps.

Detailed study of $S_{RN}1$ reactions has revealed curious features not readily accommodated by the proposed mechanism. In an alternative interpretation, the intuitively surprising suggestion has been made that the key reaction step is a bimolecular interaction between the radical anion and the nucleophile. It appears that the last word on these systems has yet to be written.

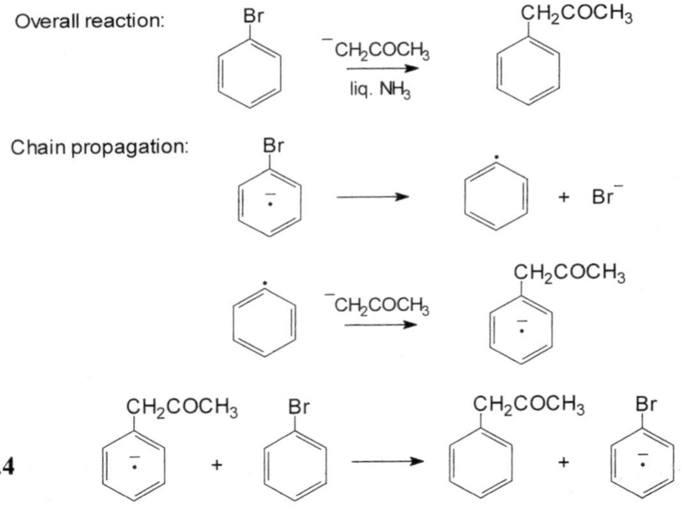

Scheme 6.4

Just as radical anions are known both as short-lived reaction intermediates and as more persistent species, the same is true of radical cations. An example of the persistent type is the isolable (blue) radical cation salt, tris-*p*-bromophenylaminium hexachloroantimonate (**6.6**), which has found a niche as a specialist oxidising agent. It reacts, for example, with *N*-vinylcarbazole in methanol to give the dimeric species **6.7** almost quantitatively.

Another dramatic application of **6.6** is as a catalyst for certain Diels-Alder cycloadditions. Thus 1,3-cyclohexadiene will dimerise slowly at 200°C to give **6.8**. But in the presence of **6.6** the reaction is rapid at −70°C. The reader will be familiar with the requirement that a 'good dienophile' in a Diels-Alder cycloaddition should incorporate a strongly electron-withdrawing substituent. Evidently, one way of achieving this is by removing one electron from a diene.

Most of the radical ions featured in this brief introduction have highly delocalised structures. It is intuitively unsurprising, therefore, that they are relatively easily formed. It is also worth emphasising that attempts to represent radical ions by means of simple valence structures may be misleading. For example, the benzophenone ketyl radical anion was represented above by a structure (**6.1**) which, quite reasonably, places the negative charge on the electronegative oxygen atom and the unpaired electron on carbonyl carbon, the latter feature implying unpaired electron delocalisation into the benzene rings, as in $Ph_3C\cdot$. A detailed theoretical analysis actually shows that only about half of the negative charge is associated with oxygen. Furthermore, there is a significant proportion (*c.* 20%) of the unpaired electron associated with the oxygen (e.g. **6.1a**) – confirmed experimentally by analysis of the e.s.r. spectrum of a sample of benzophenone ketyl which incorporated the (magnetic; $I = 5/2$) ^{17}O isotope.

6.1a

Other e.s.r. evidence is consistent with a skewing of the phenyl substituents in **6.1** (by *c.* 30°) to avoid interference between the *ortho* hydrogens (*cf.* triphenylmethyl; side-note on p. 75).

Electron delocalisation is not in all circumstances a prerequisite for facile oxidation or reduction: simple non-conjugated amines are relatively easily oxidised by removal of one of the non-bonding (lone-pair) electrons, although such reactions are seldom preparatively useful. On the other hand, amine radical cations generated from *N*-chloramines *are* synthetically important. One example of this, the Hoffmann-Löffler-Freytag reaction, was mentioned in Chapter 3 (p. 50). Conversely, the presence of low-lying vacant antibonding orbitals in, for example, non-conjugated carbonyl compounds facilitates one-electron reduction. A consequence of this is the pinacol reaction outlined above.

The ease of these SET processes is reflected in the oxidation and reduction potentials of the various substrates, and these have been investigated

Complete oxidation of hydrazine to give water and N_2 is highly exothermic. But the buttressing of the two nitrogen lone-pairs also makes one-electron oxidation of hydrazines a facile process. For example, modest concentrations of the corresponding radical cation can be prepared by oxidation of tetramethylhydrazine by **6.6** in acetonitrile:

$$
\begin{array}{c}
\text{Me} \quad\quad \text{Me} \\
\diagdown \quad\quad \diagup \\
\text{N}\!-\!\text{N} \\
\diagup \quad\quad \diagdown \\
\text{Me} \quad\quad \text{Me}
\end{array}
$$

$$\Big\downarrow (6.6)$$

$$
\left[
\begin{array}{c}
\text{Me} \quad\quad \text{Me} \\
\diagdown \quad\quad \diagup \\
\text{N}\!-\!\text{N} \\
\diagup \quad\quad \diagdown \\
\text{Me} \quad\quad \text{Me}
\end{array}
\right]^{+\cdot}
$$

In this terminology, a monoradical, with one unpaired electron, is in a 'doublet' state.

extensively by electrochemical techniques. Ionisation potentials and electron affinities may also be instructive, but these are derived from gas-phase data and necessarily take no account of solvation effects.

The activation barriers of SET reactions are generally only slightly above the heats of reaction. The additional contribution is largely a consequence of solvent reorganisation brought about by the redistribution of electric charge.

6.2 Radical pairs and biradicals

Up to this point, the emphasis has been on the chemistry of species with a single unpaired electron. Remember that any spatial atomic or molecular orbital is capable of accommodating two electrons of opposite spin. A driving force in the chemistry of radicals and radical ions is the tendency to form species with an even number of spin-paired electrons occupying bonding (or possibly non-bonding, e.g. lone-pair) orbitals. But what of situations in which two orbitals are singly occupied? We know that some species of this kind are stable. They occur in particular when two electrons are available to occupy two orbitals of equal or almost equal energy, as occurs in molecular oxygen and in many transition metal ions. In these cases, Hund's Rule applies (see also p. 57), and in the stable structure the two electrons have parallel spins, and must therefore reside in different orbitals.

When an atom or molecule with an even number of electrons has two of them with parallel spins (represented as ⇈), it is in a 'triplet' state. A molecule with all of its electrons spin-paired (⇅), irrespective of orbital occupancy, is in a 'singlet' state.

Let us now consider the possibility of two singly occupied orbitals being present in hydrocarbon molecules. It is formally possible to write open-chain structures $[\cdot CH_2(CH_2)_nCH_2\cdot]$, where n takes any value from zero upwards. Several important generalisations can be made. The unpaired electrons will interact such that in each case singlet or triplet biradicals are formally possible. If these were 'pure' spin states, interconversion would be strictly forbidden, but, especially when n is large and the two electrons are relatively far apart, the interaction is small and there is some state mixing (we may think of this as though there is 'cross-contamination' of the singlet and triplet states) so that interconversion does occur. It is only from the singlet state that electron pairing can occur to create a new σ-bond and thus form the cycloalkane, $(CH_2)_{n+2}$.

In a radical pair within a solvent cage (see p. 19), the situation is rather similar to that of $\cdot CH_2(CH_2)_nCH_2\cdot$ when n is large. Thus there is weak spin correlation between the two radical centres, and we may speak of singlet and triplet radical pairs. The interaction is small, and the energies of the two spin states are almost identical. Nevertheless, important consequences flow from the requirement that radical coupling or disproportionation can occur only from the singlet pair. We shall discuss radical pairs at the end of this Section. Before that, some individual biradicals will be introduced, beginning with $\cdot CH_2(CH_2)_nCH_2\cdot$ with n = 0, and including methylene itself, CH_2.

$\cdot CH_2CH_2\cdot$. We are dealing here with electronically excited forms of ethene (ethylene). Ethene absorbs light in the vacuum ultraviolet (λ_{max} *c.* 170 nm – equivalent to *c.* 700 kJ mol^{-1}), causing one electron to be promoted from the

π-bonding to the π*-antibonding orbital. Importantly, this is a 'vertical' process; i.e. the actual electronic transition occurs so rapidly that no adjustment takes place in the positions of the atoms. The result is an excited-state geometry which is highly unstable. In solution, excess vibrational energy may be transferred to adjacent molecules as the excited ethene relaxes to an optimum geometry in which there is a 90° dihedral angle between the two methylene groups (**6.9**). The new 'twisted' structure is some 200 kJ mol^{-1} lower in energy than the initially formed 'vertical excited state', but is still more than 450 kJ mol^{-1} above the planar ground state. In this geometry, the two unpaired electrons reside in (orthogonal) *p*-orbitals on the two carbon atoms, and the designation 1,2-biradical seems wholly appropriate. There is one further, and very important point to recognise here. It is that throughout the sequence of events just described, electron spin is conserved, so that the singlet character of the ground-state molecule is carried through to the eventual twisted excited state (or singlet biradical), **6.9**.

So how may we arrive at the *triplet* 1,2-biradical form of ethene? Hund's Rule once more applies, and the triplet is of appreciably lower energy than the singlet. But in the case of ethene 'intersystem crossing' (ISC) is very inefficient. This is usually the case when the two singly occupied orbitals can interact strongly, and the energy difference between the two states is large. The answer to accessing the triplet involves a process known as triplet sensitisation. In the case of ethene, this can be accomplished with photo-excited mercury atoms which are readily promoted into a triplet state by irradiation at 253.7 nm. These will transfer excitation energy to ethene, in a process in which electron spin is conserved (Eqn 6.1). This particular

$$^3Hg^* \ + \ CH_2{=}CH_2 \longrightarrow \ ^3\{CH_2{=}CH_2\}^* \ + \ Hg \qquad (6.1)$$

example of triplet sensitisation would be observed in the gas phase, but similar instances of energy transfer occur with larger molecules in solution. Provided that the triplet energy (i.e. the difference between the energy of the triplet excited state atom or molecule and that of the singlet ground state) of the sensitiser (in our case mercury atoms) is greater than that of the acceptor (ethene) then the reaction is normally diffusion controlled, and generates the 'vertical' triplet; i.e. for ethene, a planar species which, as with the singlet, rapidly relaxes to a preferred, twisted geometry (**6.10**).

A chemical manifestation of this twisted geometry is the equilibration of *cis*- and *trans*-1,2-dideuteroethene by photoexcitation to either the singlet or the triplet state (Eqn 6.2).

$$\text{(6.2)}$$

Once formed, triplet excited states tend to survive much longer than the corresponding singlets, since there is now a spin-conservation restriction on decay to the (singlet) ground state.

Although the case of ethene is fundamental, it is also rather unpractical because of the very large amounts of energy necessary for excitation, and the resulting complexity of the ensuing reactions. For this reason, we shall not

6.9

'Intersystem crossing' is the term used to identify interconversion of singlet and triplet states. For any molecule having an even number of electrons, two 'systems' or 'manifolds' of excited states are possible, arising from promotion of a single electron (not necessarily from the HOMO) to different antibonding MOs. Thus there is a singlet manifold of states and a triplet manifold, according to whether or not the two unpaired electrons are spin-paired. However, for almost all of organic photochemistry, interest centres exclusively on the lowest energy singlet and/or triplet state of the species concerned.

6.10

The reader may have wondered why the triplet excited state of the mercury atom is so easily generated. In the case of heavy atoms, or molecules containing them, the spin restriction is partially lifted by spin-orbit coupling.

pursue further here the actual photochemistry of ethene. The foregoing account does, however, afford a useful background to our further discussion.

·CR₂–O· and ·O–O·. Our short introduction to the photophysics of ethene prompts a brief but pertinent digression into the cases of the C=O and O=O double bonds. The first excited state of a simple carbonyl compound is much more energetically accessible than is that of ethene, requiring light of *c.* 280 nm (ca. 430 kJ mol⁻¹). But the absorption is weak, since it corresponds to a symmetry-forbidden transition in which a lone-pair (i.e. non-bonding, or 'n') electron on oxygen is promoted into a π^*-orbital. Initially, a singlet excited state is produced, but in this case, the spatial distribution of the two singly occupied orbitals is quite different. Their interaction is correspondingly smaller than in the excited state of ethene, as is the energy difference between the singlet and the triplet states, so that intersystem crossing occurs more readily. This is especially true in the case of aryl ketones, where the π^*-orbital embraces the aromatic substituent(s). For such compounds intersystem crossing is particularly rapid and photoreactions occur almost exclusively from the so-called $^3n\pi^*$ ('triplet n-to-π^*') state. This analysis has two important experimental implications, which we shall illustrate with the case of benzophenone. In the first place, the reactions of the rapidly formed triplet can adequately be described in terms of structure **6.11**. The unpaired electron which remains in a non-bonding orbital on oxygen confers the character of an alkoxyl radical. This is borne out by the marked similarity in relative rates of hydrogen abstraction from different substrates by triplet benzophenone ($^3Ph_2CO^*$) and by *t*-butoxyl radicals. The other unpaired electron is rather unreactive, as might be expected for a highly delocalised benzhydryl-like radical. Therefore this second radical centre becomes involved in the chemistry only after reaction at the oxy-radical centre. For example, with diphenylmethane, three radical coupling products are formed resulting from the sequence of reactions indicated in Scheme 6.5.

There is evidence that the rapidity of intersystem crossing in the case of benzophenone actually arises from the near perfect energy matching of the singlet excited state ($^1n\pi^*$) and a higher triplet ($^3\pi\pi^*$ – i.e. with the highest π-bonding and lowest π-anti-bonding orbitals each singly occupied) state to which the crossing occurs. The molecule then rapidly relaxes to the lower energy $^3n\pi^*$ state.

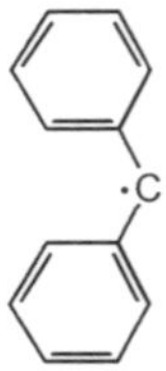

6.11

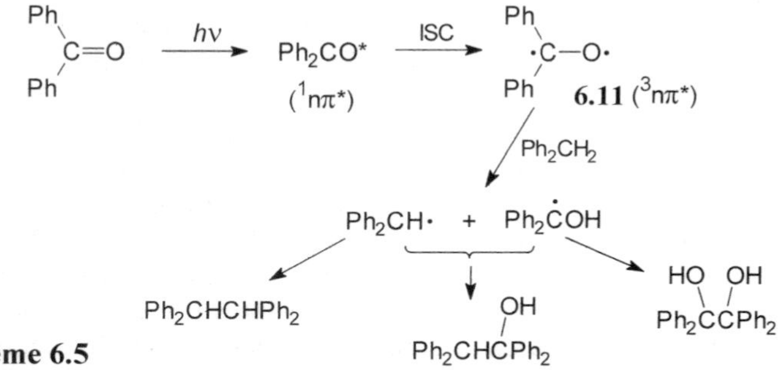

Scheme 6.5

The second point to note about benzophenone is that it is an excellent triplet sensitiser. This is because its excited triplet is so easily generated, and because it is almost as energetic as the excited singlet. Thus, although it absorbs light of longer wavelength (lower energy) than, for example, butadiene, its triplet state energy is greater than that of the hydrocarbon. Therefore irradiation (λ_{max} Ph₂CO = 345 nm; *c.* 350 kJ mol⁻¹) in the presence of butadiene is followed by diffusion-controlled triplet energy transfer (Scheme 6.6), and subsequent reactions of the triplet state of the diene.

Direct excitation of butadiene to the excited singlet takes an energy input in excess of 500 kJ mol⁻¹ (λ_{max} 217 nm), but, as with ethene, there is a large energy differerence between the excited singlet and the triplet. The triplet energy is *c.* 250 kJ mol⁻¹, whilst that of the benzophenone sensitiser is *c.* 290 kJ mol⁻¹.

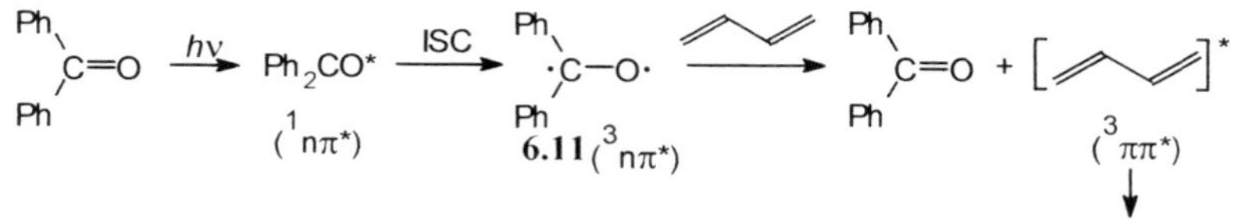

Scheme 6.6 — reaction scheme with structures:

$$Ph_2C{=}O \xrightarrow{h\nu} Ph_2CO^* \xrightarrow{ISC} Ph{\cdot}C{-}O{\cdot} \longrightarrow Ph_2C{=}O + [\triangle]^*$$

with labels $(^1n\pi^*)$, **6.11**$(^3n\pi^*)$, $(^3\pi\pi^*)$, and arrow to **Products**.

Scheme 6.6   Products

The case of molecular oxygen is intriguingly different. Here, the ground state is the triplet, and the structure drawn as O=O represents an excited state. But in this case, the energy difference between the ground and excited states is quite small (95 kJ mol^{-1}). A variety of sensitisers which absorb light in the visible region of the spectrum undergo intersystem crossing to form triplets sufficiently energetic to transfer their excitation energy to oxygen. The fact that a triplet sensitiser transfers energy to a triplet ground-state species may at first sight seem strange. But there is a significant statistical probability that spin exchange will occur in such a way that two singlet molecules will result from the interaction of two triplets. We have seen examples of the chemistry of singlet oxygen in Chapter 5. It is biologically hazardous, something which has been turned to advantage in 'photodynamic cancer therapy' in which suitable porphyrins are absorbed by tumour cells where singlet oxygen is then generated by irradiation with visible light.

CH$_2$. In the case of methylene, sometimes called 'carbene' and the parent member of the family of divalent carbon species known as carbenes, there is a new feature to be considered in the electronic structure. There are two vacant orbitals on carbon, and two electrons to be shared between them, so that there is the possibility of both singlet and triplet methylene. But now the singlet may adopt an electron configuration *in which both electrons are in the same orbital*, the second being vacant. This would have been rather unfavourable in the systems considered so far because of the energetic penalty associated with charge separation [$^+$CH$_2$(CH$_2$)$_n$CH$_2^-$]. With methylene, charge separation is not a problem, and in the most stable singlet state both electrons do indeed occupy the same orbital; the p_z-orbital (perpendicular to the H–C–H plane) is empty (**6.12**). Methylene itself, however, is a very energetic molecule, once referred to as 'the most reactive species known to man'. Generated in the singlet state, e.g. by photolysis of diazomethane, it reacts almost indiscriminately with e.g. cyclohexene. Behaving as a powerful electrophile, it not only adds across the double bond but also inserts into the *sp*^{3} C–H bonds (Scheme 6.7). In contrast, the triplet, which can be generated by photosensitised decomposition of diazomethane, behaves as a very reactive radical, with the two unpaired electrons seeming to react independently.

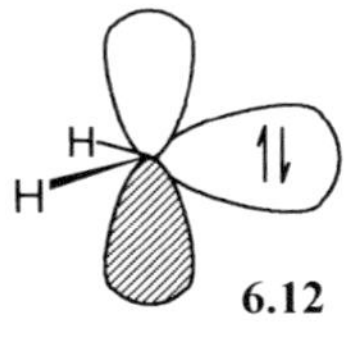

6.12

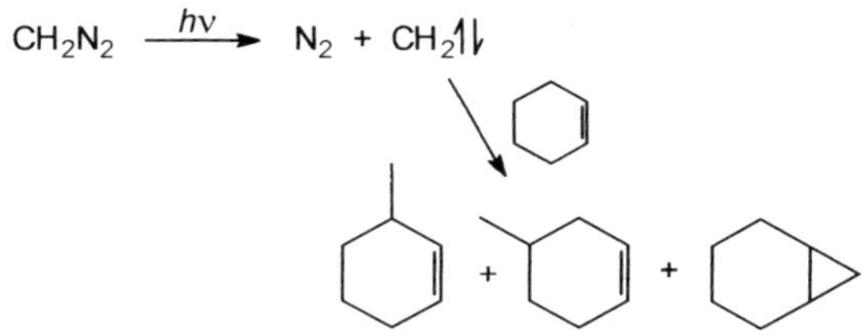

Scheme 6.7:

$$CH_2N_2 \xrightarrow{h\nu} N_2 + CH_2{\uparrow\downarrow}$$

with arrow to cyclohexene and products (cyclohexene derivatives + norcarane).

Scheme 6.7

A distinction which is often made between the reactivities of the two states of methylene is that with e.g. *cis*-butene the singlet adds to give a cyclopropane stereospecifically, driven by the interaction of the π-electrons of the alkene with the vacant orbital on methylene, whilst the triplet initially generates a *triplet* 1,3-biradical which cannot cyclise until after spin inversion has occurred. Although the interaction between the radical centres in the 1,3-biradical is small, and intersystem crossing is quite rapid, there is sufficient time for bond rotation to occur, the result being non-stereospecific cyclopropane formation. This picture, which is displayed in Scheme 6.8, is however something of an oversimplification. So much energy is released in the addition reactions that the initial vibrationally 'hot' cyclopropanes may rearrange, a complication which is especially important in the gas phase. On the other hand, less reactive carbenes, such as (singlet) dichlorocarbene and (triplet) diphenylcarbene, do in solution follow these patterns of behaviour.

Scheme 6.8

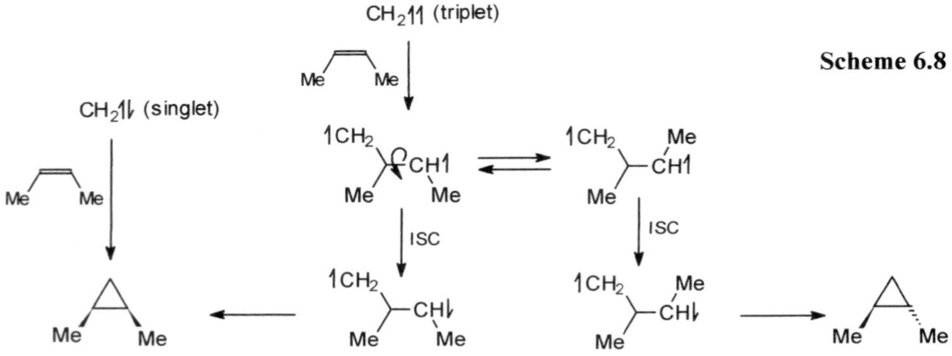

Other biradicals. One of the most widely adopted approaches to generation of 1,3- and 1,4-biradicals has been direct or triplet-sensitised photolysis of cyclic azo-compounds. These efficiently extrude nitrogen and yield products arising from intermediate alkanediyl biradicals. An instructive example is that of the *meso-* and *d,l*-isomers of the tetra-substituted tetrahydro-pyridazines, **6.13** and **6.14**. Both direct and triplet-sensitised photolysis of either isomer gives a substantial yield of 2-methyl-1-butene, together with lesser amounts of cyclobutanes. For the direct photolysis, which leads directly to singlet biradical, the stereochemistry of the precursor is largely (>90%) retained in the cyclobutane. But for the sensitised reaction, mediated by the triplet, memory of precursor stereochemistry is much lower (*c.* 60%). It is significant that in the sensitised reaction there is *some* retention of stereochemistry, since this indicates that the rate of bond rotation in the triplet biradical and that of intersystem crossing must be similar. The essential features of this chemistry are outlined in Scheme 6.9.

In other examples of formation of 1,4-biradicals, considerable variation in the rates of intersystem crossing has been observed, and it is not difficult to imagine that this will alter according to the geometric relationship between the two singly occupied orbitals. Furthermore, in some instances a wide spectrum of geometries may be explored as the conformation of the biradical

It is very important to recognise that for the successful conduct of any mechanistic investigation of triplet sensitised processes the experiments should be carried out only after careful consideration of the absorption spectra of all species involved. It should then be possible to select sensitiser and reactant concentrations, and an irradiating wavelength, such that essentially all of the light energy is absorbed by the sensitiser.

changes. An obvious example of this is the highly flexible decanediyl precursor to cyclodecane, depicted in scheme 2.17 on p. 27.

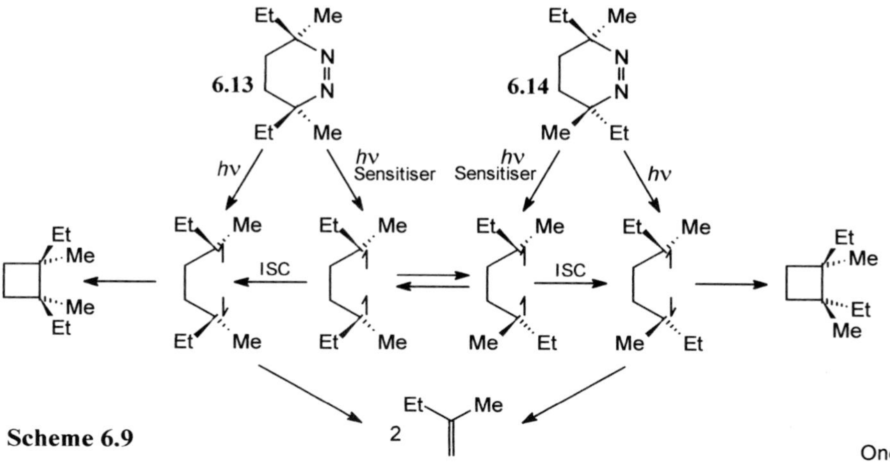

Scheme 6.9

Finally, we should take a brief look at some aromatic diyls. 1,2-Dehydrobenzene, more easily recognised as 'benzyne' (**6.17**) behaves rather as a highly strained acetylene, reacting as a powerful dienophile, and undergoing ionic addition reactions. Less familiar systems which have received attention include *meta-* and *para*-benzyne, and 1,8-dehydronaphthalene (**6. 18 - 6.20**). Of these, by far the most important is *para*-benzyne (or 1,4-dehydrobenzene; **6.19**). This is not because of any special structural features – it acts like

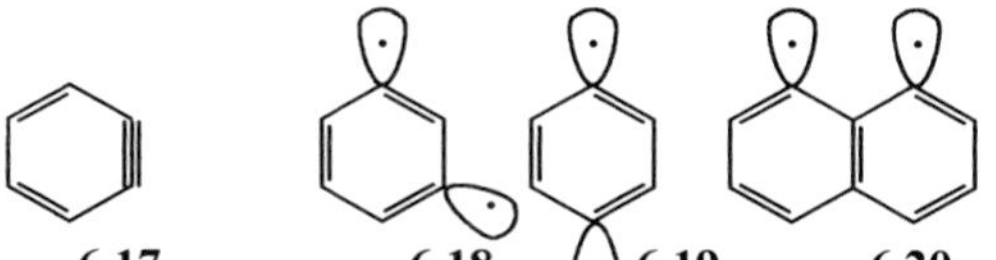

One unusual example of a 1,3-biradical which has been of great theoretical interest is that of trimethylenemethane (**6.16**). This has been produced by gas-phase photolysis of the pyrazoline, **6.15**, either with or without a triplet sensitiser. In either case, it cyclises to methylenecyclopropane.

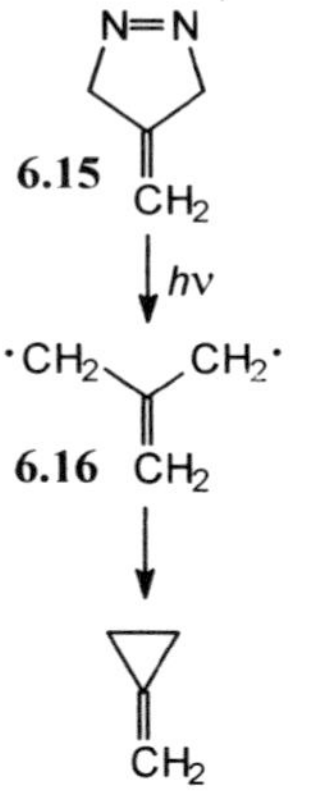

a bis-phenyl radical, with both radical centres being very reactive, and behaving independently. Its importance derives from the discovery in nature of a series of 'enediynes', powerful cytotoxic agents whose biological action depends on their ability to transform, under physiological conditions, into analogues of *para*-benzyne.

A chemical curiosity of the 1970s, named the Bergman cyclisation after its discoverer, was the observation that, when heated to *c.* 200°C, *cis*-hex-3-en-1,5-diyne (**6.21**) rearranges to give **6.19**. Evidence for the formation of **6.19** in the Bergman cyclisation included production of 1,4-dichlorobenzene when the pyrolysis was effected in the presence of CCl_4. More recently, it has been shown that, in the absence of any moderately reactive substrate, polymerisation to poly-*p*-phenylene $[-(C_6H_4)_n-]$ occurs: this unique polymer has exceptional thermal stability as well as other unusual solid-state properties.

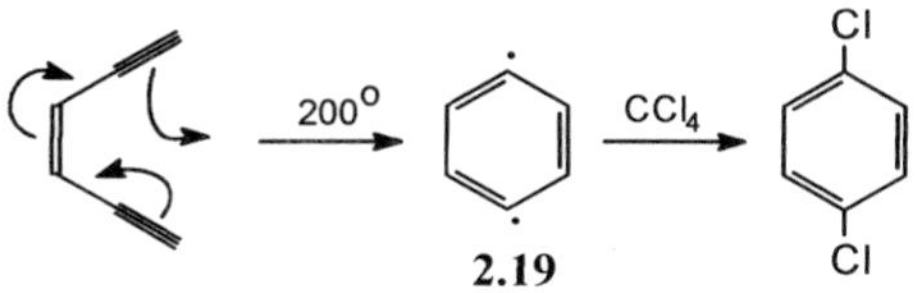

The Bergman cyclisation attracted unexpected interest in the late 1980s with the discovery *inter alia* of calicheamycin (**6.22**). What is so remarkable about this and several related natural products is their capacity to undergo the enediyne cyclisation *at ambient temperature*. This occurs however, only after the release of a trigger mechanism, which, in the case of **6.22**, involves the attack of a nucleophile on the unusual trisulphide link. Intramolecular attack of sulphur upon the carbonyl-activated double bond then produces a tricyclic system in which the ends of the enediyne moiety have been drawn close together, greatly reducing the barrier to cyclisation (see Scheme 6.10). If the nucleophile is a DNA base, the reactive biradical may react further with neighbouring DNA units in a manner which leads to strand breaking.

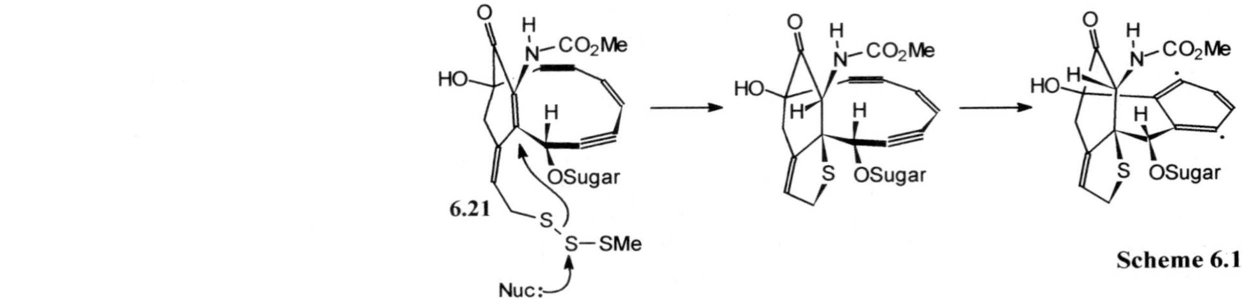

Scheme 6.10

Radical pairs. The introduction of the cage effect (p. 19), arising from recombination of geminate radical pairs, made no reference to complications arising from electron spin correlation. But thermolysis or direct photolysis of peroxide or azo-compound initiators produces singlet pairs. What happens if initiator photolysis is conducted under conditions where a triplet sensitiser absorbs the light? The answer, for example with benzophenone-sensitised decomposition of AIBN, is that the extent of cage recombination is markedly reduced. Clearly, diffusive separation must be competing with spin inversion. This is, of course, similar to the competition between bond rotation and spin inversion which was noted for 1,4-biradicals (Scheme 6.9).

Whilst 'geminate' is used for radical pairs which are formed together by initiator decomposition, the term 'encounter' pairs is sometimes used for those which arise by diffusive encounter. Here, for the first time, we meet a physical consequence of the designation triplet, for there are three possible triplet sub-states but only one singlet. The quantum statistical probability of each is equal. Therefore, to a first approximation, only one quarter of diffusive encounters should result in coupling or disproportionation. This is an approximation since it makes no allowance for intersystem crossing in the encounter radical pair.

The rate of intersystem crossing in a radical pair is influenced by the magnetic properties of the individual radicals. This includes the presence of magnetic nuclei in the radicals, including e.g. ^{1}H and ^{13}C.

Physical manifestation of these magnetic effects was first detected in two independent series of spectroscopic studies in the 1960s. In one of these, a solution of dibenzoyl peroxide in cyclohexanone was heated to 383 K in an n.m.r. spectrometer, with a view to monitoring peroxide decay by following the rise of the singlet peak due to product benzene (Eqn 6.3). But the peak behaved in a totally unexpected fashion. It rapidly grew in a *negative* sense: i.e. an n.m.r. *emission* signal was observed. This reached a maximum after about four minutes, and then decayed, eventually reverting to a normal absorption signal after the peroxide had completely decomposed.

$$(PhCO_2)_2 \longrightarrow\!\!\!\longrightarrow Ph\!\cdot\ \xrightarrow{\ RH\ }\ PhH + R\!\cdot \qquad (6.3)$$

In a later example, the thermal decomposition in benzene of the mixed diacyl peroxide **6.23** was examined in the presence of iodine. A strong transient emission signal was observed from the cage recombination product 1,1,1-trichloroethane, but now a comparably intense transient *absorption* signal ('enhanced absorption') was observed from methyl iodide, formed from methyl radicals which had escaped the solvent cage. What is happening here is that protons having nuclear spin in one direction facilitate intersystem crossing, and are therefore concentrated in the cage-escape product; protons with the opposite nuclear spin are concentrated in the cage recombination product.

The phenomenon of anomalous n.m.r. signal intensities during the course of a free-radical reaction has been designated 'Chemically Induced Dynamic Nuclear Polarisation', or CIDNP for short (pronounced 'kidnap' or sometimes 'sidnip'). As is made clear in the example of **6.23**, CIDNP phenomena are caused by a temporary disturbance (polarisation) of the normal nuclear spin populations of the observed products. CIDNP effects may be added to the list of tests diagnostic of radical participation in a chemical reaction (see Chapter 4), although a word of caution is in order. Whilst it is true that CIDNP is indicative of radical precursors to the products displaying these effects, the radical route may not necessarily be the principal reaction pathway, since as little as 1% spin polarisation can result in a 1000-fold signal enhancement. A similar cautionary note is applicable to some of the other mechanistic tests for radical intermediates, for example spin trapping (p. 57).

So far, our examples have concerned products for which the ${}^1H$ n.m.r. spectrum consists of a single line. This brief introduction to the phenomenon of CIDNP would be incomplete without mention of another type of polarisation, referred to as the 'multiplet effect'. When the diacyl peroxide/iodine experiment is repeated using dipropionyl peroxide (**6.24**), the iodoethane produced shows a transient polarisation pattern of the form sketched in Fig. 6.2. Considering the methyl triplet, which in the normal absorption spectrum would exhibit a familiar 1:2:1 triplet pattern, we see one of the outer lines of the triplet in emission and one in enhanced absorption. The centre line is rather weak. Here, if both methylene protons have the same spin they may either enhance or reduce intersystem crossing and cage escape, whilst if they have opposed spins their effects on the rate of ISC cancel. A similar analysis may be extended to the methylene quartet, which is similarly distorted from the usual 1:3:3:1 pattern.

The observable consequences of spin correlation effects in radical pairs are not confined to spectroscopic phenomena. It was recognised some years ago that magnetic (nuclear spin) differences between isotopes should influence the partitioning of radical pairs between recombination and

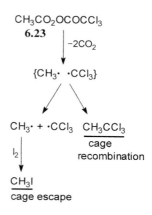

$$CH_3CO_2OCOCCl_3$$
6.23

$$\Big\downarrow -2CO_2$$

$$\{CH_3\!\cdot\ \ \cdot CCl_3\}$$

$$CH_3\!\cdot\ +\ \cdot CCl_3 \qquad CH_3CCl_3$$
cage
recombination

$$I_2\Big\downarrow$$

$$CH_3I$$
cage escape

$$(C_2H_5CO_2)_2$$
6.24

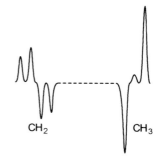

CH₂   CH₃

Fig. 6.2 A representation of the CIDNP spectrum of C₂H₅I produced during the thermal decomposition of **6.24** in the presence of iodine.

diffusive separation. Not only has this proved possible, the effect has been magnified by enhancing the pair life-time, either by micellar encapsulation, or by adsorption on silica. In one example, dibenzyl ketone with natural isotopic composition, was dispersed in an aqueous detergent solution and photolysed to 95% conversion. Unreacted ketone was then found to contain more than 3% ^{13}C at carbonyl carbon. In this system, the excited state ketone undergoes rapid intersystem crossing to give the $n\pi^*$ triplet which then dissociates to generate a triplet benzyl radical/phenylacetyl radical pair. The presence of ^{13}C at the carbonyl radical centre promotes intersystem crossing to the singlet pair, thereby facilitating recombination to the ketone.

Tailpiece

This brief text has attempted to show organic radical chemistry in a millennium setting. In other words, it has set out to paint a broad picture of the important concepts and principles, sometimes qualitatively, sometimes in greater depth, as they are understood at the end of the XXth Century. The subject continues to develop rapidly, notably in synthetic applications. But we shall conclude with two recent examples taken from the arena of the life sciences. Both involve phenoxyl radicals.

In Chapter 5, it was noted that prostaglandin synthesis was under the control of an enzyme, cyclooxygenase. A remarkable discovery has recently been made in plant biochemistry concerning the formation of the plant polymer 'lignin' (which contributes to the rigidity of woody structures) and of smaller precursor molecules called lignans. The lignans are formed in turn by oxidative coupling of phenolic monomers such as coniferyl alcohol (**6.25**).

In vitro coupling occurs principally between O-C(5'), C(5)-C(8'), and C(8)-C(8'); the products are racemic

Under protein control: exclusive C(8)-C(8') coupling also with stereocontrol

The oxidation can be reproduced in the laboratory with chemical oxidising agents and with several quite different oxidising enzymes. However, in each case the product mixture is complex, and there is no stereochemical control. The new discovery is that, unlike the prostaglandin system, in which the enzyme controls the total biochemical process, a separate glycoprotein, which is *not* an oxidation catalyst, controls the coupling of the phenoxyl radicals. When this 'dirigent' (i.e. directing) protein is present, a product distribution is obtained which is unlike that formed in its absence. The protein also controls the stereochemistry of coupling, so that a major primary product from coniferyl alcohol is the lignan, **6.26**, having the absolute stereochemistry shown.

The second example is from the challenging field of enzyme mimicry. The enzyme galactose oxidase catalyses the oxidation, by molecular oxygen, of the primary alcohol group of the sugar into aldehyde. Investigations of this enzyme had shown that the active site contains copper, and early studies had suggested that the mechanism involves the higher oxidation state, Cu(III). However, more recent work has established that the critical form of the enzyme contains copper(II) coordinated to the oxygen atom of a phenoxyl radical; its mode of action involves the abstraction of hydrogen by the phenoxyl-radical ligand. In a remarkable piece of work, the structurally related copper complex, **6.27**, has been isolated and has been shown to catalyse the oxidation of simple primary alcohols such as ethanol to the corresponding aldehyde, apparently by a mechanism which mimics that of the enzyme. The process can be cycled many times. One astonishing feature is that the hydrogen peroxide produced does not appear to poison the system. The reaction was even advocated as a means of generating high concentrations of H_2O_2 in certain organic solvents!

D-(+)-galactose

Further reading:

Although it was published almost 40 years ago, the two volume "Free Radicals", edited by J.K. Kochi and published by Wiley in 1972, still provides an excellent source of information on production, structure, and physical and chemical properties.

Any student of the history of science may be surprised by the insights already attained by W.A. Waters in his "Chemistry of Free Radicals" published in 1946 by Oxford University Press, little more than a decade after the ground-breaking work of Hey, Kharasch, Waters himself, and others during the 1930s.

More recent general surveys include:
"An Introduction to Free Radicals", by J.E. Leffler, Wiley, New York, 1993.
"Radical Chemistry", by M.J. Perkins, in the Ellis Horwood series in Organic Chemistry, Ellis Horwood, London, 1994.
"Free Radicals in Organic Chemistry", by J. Fossey, D. Lefort, and J. Sorba", Wiley, Chichester, 1995.
Of these, the first has a pronounced physicochemical flavour.

Titles highlighting applications in synthesis include:
"Radicals in Organic Synthesis: Formation of Carbon-Carbon Bonds", by B. Giese, Pergamon Press, Oxford, 1986.
"Free Radical Chain Reactions in Organic Synthesis", by D. Crich and W.B. Motherwell, Academic Press, London, 1991.

A more recent survey of radical chemistry in synthesis, emphasising efforts to achieve stereocontrol, will be found in "General Aspects of the Chemistry of Radicals", edited by Z.B. Alfassi, Wiley, Chichester, 1999. This book is one of a series of multi-author works under the general heading *The Chemistry of Free Radicals* edited by Alfassi. At the time of writing, the series also includes volumes on peroxyl radicals, sulphur radicals, and nitrogen radicals.

Specialist coverage of electron spin resonance will be found *inter alia* in:
"Theory and Applications of Electron Spin Resonance", by W. Gordy, (Techniques of Chemistry, Vol. XV), Wiley, New York, 1980.
"Principles of Electron Spin Resonance", by N.M. Atherton, Ellis Horwood, Chichester, 1993.

Index: